Design and Construction of Concrete Floors

Design and Construction of Concrete Floors

George Garber

CRC Press
Taylor & Francis Group
Boca Raton London New York

CRC Press is an imprint of the
Taylor & Francis Group, an **informa** business

Butterworth-Heinemann an imprint of Elsevier
Linacre House, Jordan Hill, Oxford OX2 8DP
30 Corporate Drive, Suite 400, Burlington, MA 01803, USA

First published in 1991 by Edward Arnold
Second edition in 2006

Copyright © 2006, Elsevier Ltd. All rights reserved

No part of this publication may be reproduced in any material form (including photocopying or storing in any medium by electronic means and whether or not transiently or incidentally to some other use of this publication) without the written permission of the copyright holder except in accordance with the provisions of the Copyright, Designs and Patents Act 1988 or under the terms of a licence issued by the Copyright Licensing Agency Ltd, 90 Tottenham Court Road, London, England W1T 4LP. Applications for the copyright holder's written permission to reproduce any part of this publication should be addressed to the publisher

Permissions may be sought directly from Elsevier's Science and Technology Rights Department in Oxford, UK: phone: (+44) (0) 1865 843830; fax: (+44) (0) 1865 853333; email: permissions@elsevier.co.uk. You may also complete your request on-line via the Elsevier homepage (http://www.elsevier.com), by selecting 'Customer Support' and then 'Obtaining Permissions'

British Library Cataloguing in Publication Data
A catalogue record for this book is available from the British Library

Library of Congress Cataloguing in Publication Data
Library of Congress Control Number: 2006925413

ISBN 13: 098-0-75-066656-5
ISBN 10: 0-75-066656-0

For information on all Butterworth-Heinemann publications visit our website at
http://books.elsevier.com

Typeset by Cepha Imaging Pvt Ltd, Bangalore, India

Working together to grow
libraries in developing countries

www.elsevier.com | www.bookaid.org | www.sabre.org

ELSEVIER   BOOK AID International   Sabre Foundation

Contents

Introduction		1
1	**Thinking about floor design**	3
	The floor's dual role	4
	A user-oriented approach	4
	Beyond structural design	5
	A balanced approach	6
	Method versus performance specifications	7
	Single-course floors	7
	Standards	8
	Lessons from roadbuilding	9
	Remedies for bad work	10
Part I The uses of concrete floors		**11**
2	**Non-industrial floors**	13
	Residential floors	14
	Office floors	17
	Floors for the retail trade	18
	Institutional floors	18
3	**Warehouse floors**	21
	Storage systems	23
	Materials-handling systems	29
	Battery-charging areas	37
	The load-class system	37

vi Contents

4   **Factory and special floors**   **41**
Factory floors   41
Special floors   45

Part II   Structural design   53

5   **Structural design of ground-supported floors**   **55**
Components   55
Structural types   58
Design methods   60
Design based on experience   65
Design considerations   66

6   **Structural design of suspended floors**   **83**
Types of suspended floors   84
Choosing a floor type   93
The design process   93
Design considerations   95

Part III   Concrete for floors   105

7   **Properties of plastic concrete**   **107**
Workability   107
Finishability   117
Bleeding   120
Setting time   122
Plastic settlement   124

8   **Properties of hardened concrete**   **127**
Strength   127
Impact resistance   138
Modulus of elasticity   139
Drying shrinkage   140
Thermal coefficient   142

9   **The components of concrete**   **145**
Cement   145
Aggregates   152
Water   159
Admixtures   160
Fibres   167

10	**Mix design and mixing**	**169**
	Mix design	169
	Mixing	172
11	**Transporting and placing concrete**	**175**
	Transporting concrete	175
	Slab layout	181
	Side forms	190
	Tools for placing concrete	191
	Compaction	193
	The next step	194
12	**Curing**	**195**
	Curing methods	196
	How to choose a curing method	203
	Timing	203

Part IV Joints and Cracks — 205

13	**Cracks**	**207**
	Plastic-shrinkage cracks	207
	Plastic-settlement cracks	208
	Crazing	209
	Drying-shrinkage cracks	210
	Thermal-contraction cracks	211
	Structural cracks	212
	Crack repair	214
14	**Curling**	**219**
	False curling	221
	Resisting curl	221
	Limiting curl	222
	Designing around curl	223
	Repairing curled slabs	224
15	**Joints**	**227**
	The function of joints	227
	Joint types	228
	Load transfer at joints	233
	Inducing joints	241
	Joint fillers	245
	Armoured joints	249
	Joint sealants	249

Contents

16 Crack control in ground-supported floors — 253
- The "let it crack" approach — 254
- Unreinforced floors with joints — 254
- Reinforced floors without joints — 257
- Reinforced floors with joints — 262
- Prestress — 267
- Sub-slab friction — 274

Part V The floor surface — 275

17 Floor finishing — 277
- Principles of finishing — 277
- Tools for finishing — 278
- The order of finishing steps — 290
- Types of finishes — 291
- Other finishing methods — 294
- How to specify finishes — 297

18 Concrete toppings — 299
- Monolithic toppings — 299
- Bonded toppings — 301
- Unbonded toppings — 302
- Forbidden thicknesses — 303
- Cement–sand screeds — 303
- Terrazzo — 309
- Summary – choosing a topping — 310

19 Surface regularity — 311
- Flatness and levelness — 312
- Defined versus random traffic — 312
- F-numbers — 313
- The TR 34 system — 320
- Straightedge tolerances — 325
- Factors that affect surface regularity — 326
- Superflat floors — 329

20 Resistance to wear — 333
- The traditional approach — 334
- Classifying wear resistance — 335
- Testing wear resistance — 337
- How to specify wear resistance — 340
- Factors that affect wear resistance — 340
- Improving wear resistance — 345

21	**Resistance to chemical attack**	**349**
	Concrete's chemical resistance	349
	Protecting the concrete	350
	Attack from below	353
22	**Preparation for coatings, toppings and floorcoverings**	**355**
	Moisture	355
	Bond	361
	Surface regularity	362

References 365

Glossary 369

Index 379

Introduction

1

Thinking about Floor Design

Concrete floors lie all around us. Every building has a floor, and in most industrial and commercial buildings that floor is made of concrete. Many houses also have concrete floors.

Regrettably, many of those floors fail to do their job. Floors are responsible for more user complaints than any other building element except roofs.

Good floors exist, but they seldom appear by accident. They require good design, which requires a philosophy, by which I mean a coherent, integrated way of looking at a subject. Too often, a floor design is a jumble: a list of national standards, some clauses borrowed from the last project manual and a few heavily promoted, proprietary products. That is no way to do the job.

This chapter sets out the philosophy on which the rest of the book is based. The essential ideas are these:

- A floor has a dual role. It is part of a building, but it is also part of the building user's equipment.
- A floor must be designed around the user's needs and wishes.
- Floor designers should not focus narrowly on structural strength, but must consider other important properties of the floor.
- Good design requires equal attention to five factors: floor usage; structural strength; properties of the concrete itself; cracks and joints; and properties of the floor surface.
- Performance specifications usually work better than method specifications.
- Single-course floors usually work better than double-course floors.
- National standards should be used with care.
- Lessons from roadbuilding should be applied to floor construction.
- Remedies for bad work should be specified before the bad work occurs.

Though this book is inevitably based on my own ideas, it is more, I hope, than a description of my own personal approach to floor design. I have tried to present both sides of controversial issues, of which the concrete-floor industry has many. If I have failed to address a widely used idea or product, the reason is my ignorance, not my opposition to it.

The floor's dual role

More than any other building element, floors exist in two different worlds.

On one hand, a floor is part of a building's structure. It is designed by a structural engineer, who may rely on it to support or restrain other parts of the building. It is built under a construction contract. It is bought and sold with the building, and is usually expected to last the life of the building.

On the other hand, a floor is part of the user's equipment. It may be a work surface on which the user assembles products, a roadway for warehouse trucks, or a platform for storing goods.

This dual role sets floors apart from other building elements. A roof, for example, does the same job on every building. Whether it covers a garden shed or a steel mill, its task is to keep out rain and snow. In contrast, floors do vastly different jobs in different buildings. The floor that works in a house would never do in a dairy – and vice versa.

We who design and build floors spend most of working hours in the world of building construction. We may not often think about the floor users and their needs – but we should. To those users, the floor is far more than just another building element. It is something they use and live with every day.

A user-oriented approach

Because floor uses differ widely, every floor should be designed with the user's needs in mind. Almost no one would disagree with that advice, but how many designers really follow it?

In too many cases, what passes as user-oriented design is little more than finding out the floor loadings and designing the floor to support them structurally. That is an essential step, but it is only part of the job.

Floor users have a long list of complaints, and most have nothing to do with structural failure. On some floors, joints break down and must be repaired again and again. Other floors become unusable when surfaces wear out. Huge warehouses have been shut down for repairs because floors were not flat enough for a particular vehicle. On non-industrial floors, complaints about moisture probably head the list. On industrial floors, joint problem predominate.

We can prevent most problems when we design and build floors with users' needs in mind. The process starts with asking users what they need and want, but it does not end there. Users – and where appropriate, their equipment suppliers – should take part in many design decisions, particularly when considering costs over the life of the building. Some users with little cash will accept higher maintenance costs in exchange for the cheapest possible new floor. Other users prefer to pay a little more today to reduce future troubles. Some users will gladly pay a lot more today. Any one of those choices may make sense, depending on the user's circumstances and desires.

Besides producing floors that are better fitted to their purpose, a user-oriented approach often makes floors cheaper as well. When a floor is designed to do one specific job, the designers can eliminate features that would be needed in a multi-purpose floor. The savings can be large.

The one case where a user-oriented approach would seem impossible is the speculative warehouse, put up in the hope of attracting a tenant or buyer. The floor user could be almost anyone, doing almost anything. But even there a solution exists. Some property developers build speculative warehouses without floors. They install a level, granular base so prospective users can walk around in their dress shoes, but they leave the floor out. Once a contract has been signed, they design and build the floor to suit the tenant or buyer, whose exact needs can then be determined.

The user-oriented approach to floor design is becoming more important as users demand more of their floors. The growing popularity of high-stack, very-narrow-aisle warehousing and flexible manufacturing is making special floors not only desirable but essential.

Beyond structural design

Related to this user-oriented approach is the need to go beyond structural strength in designing floors.

Most commercial and industrial floors are designed by structural engineers. Not surprisingly, such engineers look first to the structural strength of the floor. There is nothing wrong with that, as long as they go on to consider other qualities. But because of their training and experience, structural engineers may pay too much attention to the floor's role as part of the building's structure, and too little attention to the specific needs of the floor user in areas other than structural strength.

In the past, many floor designers focused too narrowly on a floor's structural strength. They examined floor loads and carefully calculated the structural design. But they spent little time on other important factors such as joint details, wear resistance and flatness. When they tried to control such factors by specification, they often relied on few shopworn phrases that had little effect on the final product.

That was never a good way to design floors, but it may have been tolerable in the days of hand-pushed trolleys. In today's world of high-speed forklifts, robot vehicles and storage racks 30 m high, it simply will not do.

I do not advocate that structural design be ignored (and there is no chance that will happen, anyway). Nor do I mean to say that structural engineers should not design floors; many of them do a good job. But we need to see structural analysis as just one part of the design process, and structural engineers must remember that, to design good floors, they have to go beyond structural design.

In this book I pay relatively little attention to structural design, for three reasons. First, I know that writers more qualified than I have already covered the subject. Second, a thorough treatment of structural design would have required a much longer and different book. And third, I want to offset the traditional overemphasis on structural analysis.

A balanced approach

Instead of concentrating on structural strength, we need to consider all the properties that make a floor work well. This book is divided into five parts, each of which deals with an equally important aspect of floor design.

Part I discusses the uses of concrete floors. We cannot even begin to design good floors until we understand how people use them.

Part II deals with structural design. While it is not a complete design manual, it discusses the structural types, presents the main issues to be addressed, and identifies some of the more useful design guides, which vary from country to country.

Part III discusses the concrete used for floors. Floor construction makes unusual demands of concrete. Concrete that works perfectly well on footings and walls may be wrong for floors.

Part IV deals with cracks and joints. Concrete wants to crack, but most people prefer that it remain uncracked. For that reason crack control is an important part of almost every floor design, but the methods for controlling cracks vary greatly. Joints are included here because they are, in many cases, the floor's chief defence against cracks.

Part V discusses properties of the floor surface. They include surface finish, flatness and levelness, wear resistance and resistance to chemical attack. Part V also deals with toppings and floorcoverings.

Method versus performance specifications

A method specification tells the builder how to build something. A performance specification describes a property that the finished product must possess, without telling the builder exactly how to achieve that property.

While most floor designs include both kinds of specifications, this book is biased toward performance specifications. Well-written performance specifications, if enforced, provide the best guarantee that users will get the floors they need. Another benefit is that they encourage the development of new products and techniques.

That is not to say method specifications should never be used. They have their place on small jobs and on the less important parts of big jobs, where the cost of supervision and testing cannot be justified. They are also valuable for certain details for which no standard tests exist. But those are exceptions. This book leans toward performance specifications in almost every case.

Single-course floors

In the past, many floors had two layers: a structural slab plus a topping. Two-course floors are still built (see Chapter 18), but they are less popular than they used to be – for good reason.

Single-course floors have important advantages. They are cheaper and take less time to lay. And they eliminate the risk of bonding failure between base slab and topping. (Not every two-course floor relies on a bonded topping that can fail, but many do.)

Designers should always think first of a single-course floor, and should add a second course only when there is a compelling reason to do so. With today's methods of placing and finishing concrete, single-course floors can satisfy almost all requirements.

Standards

It is hard to imagine designing a floor without recourse to published standards. They serve two purposes, providing a common language and saving labour. Without standards, specifications and notes on drawings would have to be much longer than they are now.

We can, however, misuse and overuse standards. We need to recognize their limitations, of which there are at least three. First, standards are written for ordinary circumstances. Few if any apply to every floor. Floors for special purposes often lie outside the scope of existing standards.

Second, standards are conservative. It normally takes years for a standard-writing committee to adopt a new idea. For that reason, standards do not always incorporate the best, latest technology.

Third, standards cannot replace the designer's own analysis and judgement. As the British Standards Institution warns, "Compliance with the British Standard does not of itself confer immunity from legal obligations". It is the designer's job to decide where and how to use standards. Along with that goes the obligation to know, to at least some extent, what those standards actually say. We all sometimes cite standards we have not read, but it's nothing to brag about. I almost never fail to find a surprise or two when I read or re-read a standard.

This book cites mainly British and American standards. That's not as limiting as it sounds, because those standards have become the models for much of the world. British standards are widely used throughout the Commonwealth (though not in Canada), while American standards dominate North and South America.

Both the British and American systems can be confusing, though for different reasons. In Britain the confusion comes from the shift to European rules. To a large and increasing extent, so-called British standards are really just the English-language versions of European standards, meant to apply throughout the European Union. Many venerable British standards, such as BS 12 on cement and BS 1881 on concrete testing, have vanished. Their replacements are not necessarily better on their inherent merits, but are needed for uniformity across Europe.

In America the confusion comes from the huge number of separate standards and from the existence of two standards-writing organizations. The American Concrete Institute (ACI) controls the standards that tell us how to use concrete in construction. ASTM (formerly American Society for Testing and Materials, but now just known by the four letters) deals with test methods and specifications for materials.

British standards start with the letters BS. If a number follows right behind those letters, the standard is one of the older, strictly British documents.

If the letters EN (for Euronorm) appear after BS, the standard is European. Both kinds are available from:

British Standards Institution
389 Cheswick High Road
London W4 4AL
United Kingdom
Telephone +44 20 8996 9000

American concrete standards start with the letters ACI or ASTM. ACI standards are available from:

American Concrete Institute (also known as ACI International)
P.O. Box 9094
Farmington Hills, Michigan 48333
USA
Telephone +1 248 848 3700

ASTM standards come from:

ASTM
100 Barr Harbor Drive
West Conshohocken, Pennsylvania 19428
USA
Telephone +1 610 832 9611

Lessons from roadbuilding

Many valuable ideas for floors stem from road construction. The now-common practice of casting concrete slabs in long strips was borrowed straight from roadbuilding. The Somero Laser Screed, widely used to place concrete on big industrial floors, was invented in the 1980s as an adaptation of the big concrete paving machines used on highways. Most methods for transferring load across joints originated in highway work.

Roads and floors have much in common. Both support traffic and are more or less horizontal. Concrete highways resemble ground-supported concrete floors. Concrete bridge decks resemble suspended concrete floors.

For several reasons, roadbuilding technology is often in the lead. Highway engineers are no more skilled than floor designers, but they get more money for research. The large scale of highway work makes experimentation easier to justify and arrange. And most roads are built for governments, who encourage the publication of reports and sharing of knowledge. In contrast, most floors are built privately by firms that often wish to guard what they see as trade secrets.

The flow of knowledge from roadbuilding to floorlaying has not ended. This book presents several ideas that are well accepted in the highway industry, but not yet widely adopted for floors. In Chapter 16, I discuss the use of unreinforced slabs with closely spaced, dowelled joints. That is a standard design for concrete highways, but is still controversial in floor construction.

Not every development in roadbuilding can be applied to floors, but many can. Floor designers and builders should keep their eyes open for new ideas from this direction.

Remedies for bad work

In a perfect world, every floor would be laid exactly as specified. In the real world, people make mistakes. What happens then?

Many specifications say nothing about this subject. When problems occur, there is no clear course of action. The designer may try to reject a whole floor for one or two minor flaws. The builder will want to repair the floor as cheaply as possible, or may argue that it is usable despite its defects. Sometimes the two parties reach a compromise, which may or may not meet the user's needs. At other times there is no meeting of minds, and the matter goes to court or arbitration.

Many controversies can be prevented by deciding on the remedy before the problem occurs. For extreme problems, the remedy may be to replace the floor. But full replacement is seldom justified; it is not only costly, but often disastrous to the construction schedule. In most cases, floor defects can be repaired.

Wherever possible, specifications should state exactly what will happen when a failure is detected. This simple step can provide several benefits:

- Jobs run more smoothly, with fewer arguments.
- Builders' risks are limited, and that leads to lower costs.
- Costly litigation is reduced.

Part I

The Uses of Concrete Floors

2

Non-Industrial Floors

This category includes all floors except warehouse and factory floors, and a few special types covered in Chapter 4. Though non-industrial floors get little attention in the technical press, they are actually far more numerous than their industrial counterparts.

Non-industrial floors are usually easier to design and build than industrial floors, for three reasons. First, they generally support very light loads. Second, few of these floors are subject to wheeled traffic. Third – and this may be the most important reason – most non-industrial concrete floors are concealed beneath floorcoverings such as carpet, tile or resilient flooring.

It can be a serious mistake, however, to take these floors for granted. Problems occur even on routine house slabs. And some non-industrial floor users have special requirements that can be as demanding as those in heavy industry. For example, some libraries use movable, motorized shelves that require an unusually level floor surface.

And not every non-industrial floor stays hidden under floorcoverings. Nowadays many retail shops have exposed concrete floors, and their owners prefer a smooth, crack-free surface. Some architects specify exposed concrete slabs, often made of coloured concrete or having special decorative finishes, in public buildings and even in houses.

There is no generally accepted classification for non-industrial floors. I divide them into the following categories:

- residential floors;
- office floors;
- floors for retail trade;
- institutional floors.

Residential floors

These are found in private houses and multiple dwellings. They are characterized by short spans, light loads, and foot traffic. Nearly all residential floors are concealed by carpet, tile, sheet material or wood.

This is the only category of floor use in which concrete faces serious competition as a structural material. Suspended timber floors are common in houses, and outnumber concrete floors in many countries.

Loading

Residential floors support a variety of loads from occupants, furniture, partition walls and appliances. Most loads are very light compared to those found in industrial buildings.

Designers seldom even consider loads when planning ground-supported residential floors. Such floors are usually designed to a standard thickness without regard to specific loading. A 100 mm (4 in) slab is usual, and will suffice on all but the worst sites.

Suspended residential floors are often designed for a nominal live load of about 2.0 kN/m^2 (40 lb/ft^2). See Table 2.1. Local building regulations may require a different figure.

Occasionally a residential floor has to support a heavy load such as a masonry fireplace. Then the floor, whether ground-supported or suspended, should be designed for that particular load. It is often easier to provide a separate foundation for the heavy object.

Joint and crack requirements

Joints and cracks seldom create much trouble in residential construction, because the concrete slabs are usually concealed. Stress-relief joints are seldom needed, except for isolation joints to separate a ground-supported floor from fixed objects. Construction joints may be used, but many residential floors are small enough to be laid without them.

Most residential floors are covered, but there are a few exceptions – in garages, for example – where the concrete slab is exposed to view. In such situations the designer may wish to use stress-relief joints to reduce the risk of unsightly cracks.

Cracks in a residential floor rarely need sealing or repair unless they are so wide that they would show through a floorcovering.

Non-Industrial Floors

Table 2.1 Typical live loads for residential, office and institutional floors

Floor Use	Live Load kN/m²	lb/ft²
Residential		
Private quarters	2.0	40
Public corridors and lobbies	3.0	60
Offices		
Ordinary offices	4.0	80
File rooms	6.0	125
Schools		
Classrooms	2.0	40
Corridors	5.0	100
Libraries		
Reading rooms	3.0	60
Stacks	7.0	150
Hospitals		
Wards and private rooms	2.0	40
Operating theaters	3.0	60
Corridors	4.0	80
Prisons		
Cells	2.0	40
Corridors	4.0	80

Note: Local building regulations may differ from the figures shown here.

Surface requirements

Residential use demands little of the concrete surface. Within living quarters, the concrete is almost always concealed beneath a floorcovering or an applied flooring, which greatly reduces the demands on the concrete itself.

Surface regularity (flatness and levelness) matters little. Most residential users will be quite happy with a floor that measures at least F_F15 for flatness and F_L10 for levelness. This is a low standard, roughly equivalent to an allowable gap of 20 mm (13/16 in) under a 3-m (10-ft) straight edge. See Chapter 19 for more information on surface regularity.

Wear resistance is of no concern unless the concrete slab will be exposed to wear in the finished building. In that case, the surface should meet wear-resistance class AR3. See Chapter 18 for more information on wear resistance.

The surface finish or texture may or may not be important, depending on how the floor will be covered. Tile, carpet or wood flooring can be laid over almost any concrete finish. But thin sheet materials require a smooth surface with no obvious ripples or ridges. A good trowel finish is suitable. See Chapter 17 for more information on floor finishes.

In some countries builders normally lay a cement–sand screed over the concrete slab, before installing the floorcoverings. In view of the fact that other countries do quite well without screeds, designers should think twice before specifying screeds. The floor will almost always cost less if the screed is left out.

A different finish may be required on concrete slabs that are exposed to wear – as in garages and utility rooms. Some builders put a float finish on such floors, but most users appreciate a smooth trowel finish because it is easier to clean. It may be desirable to slope the surface for drainage.

Special considerations

Residential floors are usually simple to design and build, but there are a few special issues to consider, including:

- damp-proofing;
- supervision;
- poor soil.

Damp-proofing is very important in residential floors at ground level. Most residential floorcoverings are easily damaged by rising damp. Even if the floorcovering is not highly vulnerable, the user may change floorcoverings later. See Chapter 22 for more on this subject.

Residential floors usually get less *supervision*, inspection and testing than other floors. Because of this, designs should be as simple and foolproof as possible. It is wise to avoid procedures that require close supervision. For example, wet-curing (see Chapter 12) is seldom a good choice for residential construction because it needs near-constant attention for several days. A curing compound is safer.

Residential floors are laid on a wide variety of building sites, including many with *poor soil*. People regularly build houses on ground that would be avoided for industrial construction. Unfortunately, some effective techniques for using poor soil – cement-stabilization, vibro-compaction and piling, to name three – are hard to justify economically in small-scale residential work. In the Gulf Coast region of the USA, some builders rely on post-tensioning to improve the performance of slabs on questionable soil (see Chapter 16).

Office floors

With a few exceptions, office floors are similar to residential floors, though they sometimes support heavier loads. Most of the preceding discussion of residential floors applies equally to office floors.

Loading

Office floors are typically designed to support live loads of about 4.0 kN/m^2 (80 lb/ft^2) (see Table 2.1). Heavier loads may be found in file rooms.

Joint and crack requirements

Office floors have no special requirements for joint and cracks.

Surface requirements

Office usage demands little of the concrete floor surface. Office floors are almost invariably concealed by carpet or thin sheet material.

Surface regularity is generally unimportant, but there are some exceptions. One authority (Face, 1987) says movable partitions, used in many offices, benefit from floor levelness of $F_L 20$ or better. That degree of levelness is hard to achieve on a suspended floor, where it may require a topping.

As in residential construction, floors to be covered with thin sheet material should be smooth enough to prevent defects showing through.

Floors for the retail trade

This category covers a wide range of buildings, from small local shops to hypermarkets to do-it-yourself stores.

Floors in small and medium-sized shops are usually concealed. They are similar to office floors and are designed and built in the same way. At the other extreme, floors in big do-it-yourself stores are functionally the same as warehouse floors and should be designed as such.

Loading

Retail floors often support heavier loads than residential and office floors. The user may be able to provide specific information. Lacking that, designers typically assume a live load of about 7.0 kN/m² (150 lb/ft²). Certain trades impose much lighter loads, but it is unwise to base the design on them because future occupants may be in a completely different line of business.

Surface requirements

These depend on whether the floor gets a covering. Covered retail floors have requirements similar to those for residential and office floors. They need to be smooth enough so that defects do not show through the floorcoverings. Exposed retail floors have requirements similar to those for warehouse floors, discussed in Chapter 3. They usually get a smooth, power-trowelled finish.

Designers of retail floors seldom pay much attention to flatness and levelness, but exceptions exist. Some retail chains in America demand unusually flat floors, with flatness numbers of $F_F 35$ or even $F_F 50$. In the UK, BS 8204 says an SR2 surface, with a maximum gap of 5 mm under a 3 m straightedge, is suitable for "normal commercial floors". Almost no one enforces that, however.

Institutional floors

This category includes floors in schools, libraries, hospitals and other public buildings. These floors are characterized by heavy foot traffic, but

Non-Industrial Floors 19

may have little else in common. Many institutional floors are concealed, but direct-finished concrete is also used.

Loading

Table 2.1 shows typical design loads for some types of institutional floors. Local building regulations usually dictate the loads for which these floors are designed.

Joint and crack requirements

If an institutional floor will be covered, the joint and crack requirements are the same as for residential and office floors.

Some institutional floors are made of direct-finished concrete, however. Cracks in such floors rarely affect usability, but may nevertheless be highly objectionable to the users. For that reason, crack control may be very important – even more important than in the typical industrial floor.

Special considerations

Some institutional floors, particularly those in public buildings such as courthouses, railway stations and airport terminals, must combine structural strength, wear resistance and architectural effect. That is an unusual combination in a concrete floor, but it can be achieved.

There is no single solution to the problem, because much depends on how the designer wants the floor to look. Many designers use terrazzo toppings for these floors. Other designers have used coloured concrete, concrete with stamped patterns or stains.

3

Warehouse Floors

A warehouse is an industrial building used for storage. Not so many years ago, warehouses were places where men stacked bags and boxes by hand. Such buildings still exist, but many modern warehouses are far different. Today's warehouse might contain automatic cranes serving racks 25 m (85 ft) high. Or it might include a mix of multi-level conveyors and wire-guided turret trucks in aisles just 2 m (6 ft) wide. Even the term *warehouse* has become dated in some circles. Many industrial engineers speak instead of "materials-handling systems" and "distribution centres".

Though they vary widely, warehouse floors are often characterized by heavy loads, wheeled traffic and large scale. In some clad-rack warehouses, the floor is designed to support a concentrated load of more than 25 t (55 000 lb or 27.5 US tons) at each rack-leg. Almost every warehouse includes some vehicles that travel on the floor and control aspects of the floor design (see Figure 3.1). And as for scale – some warehouses have over 100 000 m^2 (1 000 000 ft^2) of floor at ground level.

Most warehouse floors are ground-supported. Ground-level suspended floors, built on piles, are sometimes used where the subgrade is too weak for a slab. Few warehouses floors are built as upper-storey suspended slabs, but this method of construction is sometimes adopted where land costs are very high. Multi-storey warehouses are more common in Hong Kong and Singapore than in Europe and America. Some warehouses in which the main floor is ground-supported include one or more suspended floors on mezzanines. The mezzanine floors usually support only light loads.

Direct-finished concrete provides the wearing surface in most modern warehouses. Some warehouse floors get toppings or coatings, but almost none nowadays receive floorcoverings.

Figure 3.1 A warehouse may appear to be a simple building, but its floor requirements can be complex

Depending on usage, a warehouse floor may have to meet strict requirements for surface regularity and wear resistance. Joint details and crack control are important issues in any warehouse that supports wheeled vehicles, and almost all fall into that category. Resistance to chemical attack is relatively unimportant, with the limited but significant exception of battery-charging areas, where sulphuric acid spills occur. Some designers specify epoxy coatings there.

Every warehouse contains at least one storage system and one materials-handling system, and some buildings contain more than one of each. These systems impose different demands on the floor. We often find that the storage system controls the slab thickness, while the materials-handling system dictates floor flatness and wear resistance.

In this chapter we shall look at the main kinds of storage and materials-handling systems, considering floor requirements in the following areas:

- loading;
- joints and cracks;
- surface regularity;
- wear resistance.

At the end of the chapter we shall look at ways to design warehouse floors when we do not know the exact usage.

Storage systems

These can be divided into six types:
- bulk storage;
- block stacking;
- pallet racks;
- shelving;
- mezzanines;
- mobile racks.

Bulk storage

This consists of loose material, such as corn or gravel, stored directly on the floor. Loading is almost perfectly uniform, with no concentrated point loads and no unloaded aisles. For that reason bulk storage creates little bending stress in the floor slab, even where the loads are heavy.

Here are the basic floor requirements for bulk storage:
- loading: varies greatly, but is always easy to calculate since all you need to know is the density and height of the material being stored;
- joints and cracks: may need to be sealed if the warehouse holds foodstuffs;
- surface regularity: not critical;
- wear resistance: not critical.

Block stacking

This consists of unit loads, such as goods on pallets or rolls of paper, stacked directly on the floor. The unit loads are stacked in blocks, sometimes several units deep, with aisles for vehicle access. Loads are limited by stacking height, which rarely exceeds 9 m (30 ft), even if the building is taller than that.

Block-stacked loads are usually described as if they were uniformly distributed, expressed in kN/m^2 (lb/ft^2). Loads can run as high as 100 kN/m^2 (2100 lb/ft^2), but are usually much less.

For structural design, these uniformly distributed loads are treated differently from the concentrated loads imposed by most other storage systems. With block stacking, the sub-base cannot be considered to add anything to the basic properties of the subgrade. See Chapter 5 for more information.

Here are the basic floor requirements for block stacking:

- loading: varies, but typically about 25 kN/m² (500 lb/ft²);
- joints and cracks: not critical;
- surface regularity: not critical;
- wear resistance: not critical.

Pallet racks

These are steel frames that support palletized loads. The standard configuration consists of horizontal beams spanning between vertical end frames. The beams are usually long enough to support two pallets side-by-side; some support three pallets between end frames.

Pallet racks impose heavy concentrated loads on the floor. In the standard configuration, each rack-leg (except at the end of the racks) supports

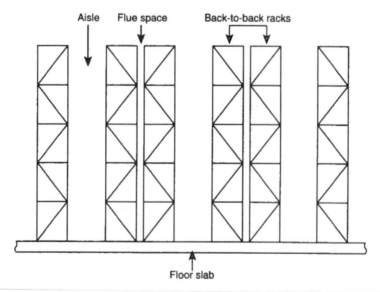

Figure 3.2 Pallet racks in the common back-to-back configuration

a weight equal to one pallet load for each level of pallets, plus the dead weight of the racks themselves. (The end rack-legs support only half as much.) The load is typically 2–8 t (4400–18 000 lb, or 2.2–9 US tons) per rack-leg. Very tall racks can produce heavier loads.

Extremely heavy loading can occur in a clad-rack building (called rack-supported in America) where wind loads add to the normal gravity loads. A few rack-supported warehouses are designed for loads of more than 25 t (55 000 lb or 27.5 US tons) at each rack-leg.

In most cases, pallet racks are erected in back-to-back pairs, separated by a narrow flue space (see Figure 3.2). This pattern puts two loaded rack-legs about 300 mm (12 in) apart. This double concentration of load often controls the floor's structural design. With push-back racks, pallet racks can be erected in blocks more than two rows wide.

The loads from pallet racks may require a very thick – and therefore costly – floor slab. To reduce slab thickness and cost, some designers specify broader baseplates for the rack-legs. This is sound engineering – but only if the baseplate is stout enough to spread the load. The typical baseplate is made of thin steel and has little load-spreading ability. Thick baseplates are available, but you have to ask for them.

In some cases the designer can reduce slab thickness substantially just by applying a sharp pencil to the estimated rack loads. Many loads are overestimated. For example, suppose we are designing the floor for a rack system four pallets high, with all pallets supported by the racks. We ask the user the weight of the heaviest pallet load, and are told it is 1.5 t (3300 lb). We then design the floor for rack-leg load of 6.0 t (13 200 lb), which is the pallet load four times four (for the four levels).

That would be a safe design, but almost certainly too conservative. We based the design on the maximum pallet load. But the average pallet almost always weighs less than the maximum – sometimes less than half as much. It makes no sense to design for a rack full of maximum loads when there is no possibility of that situation ever occurring.

A good materials-handling consultant can advise on the probability of certain load levels being exceeded in a particular warehouse.

Here are the floor requirements for pallet racks:

- loading: usually 2–8 t (4400–18 000 lb, or 2.2–9 US tons) per rack-leg, but sometimes much higher;
- joints and cracks: no need for sealing, but load transfer needs to be considered in structural design;
- surface regularity: not critical, but a level floor will reduce the need for shims;
- wear resistance: not critical.

Shelving

This is similar to pallet racking, but on a smaller scale for storing non-palletized goods. Shelving is seldom as tall or as heavily loaded as pallet racks, though it can be (see Figure 3.3). It is more demanding in one respect, however. The legs are closely spaced, with the back-to-back legs often in direct contact. The structural design must take into account this proximity of loads.

Here are the floor requirements for shelving:

- loading: usually less than 3 t (6600 lb or 3.3 US tons) per leg;
- joints and cracks: no need for sealing, but load transfer needs to be considered in structural design;
- surface regularity: not critical;
- wear resistance: not critical.

Figure 3.3 Heavy-duty shelving in a very-narrow-aisle (VNA) warehouse

Mezzanines

These are raised platforms supported by widely spaced posts (see Figure 3.4). With mezzanine storage there are two floors to consider: the suspended mezzanine floor, and the main floor, almost always ground-supported, on which the mezzanine rests.

Mezzanine floors are usually made of steel, but suspended concrete slabs are also used. They are typically designed for a nominal loading of 3–10 kN/m² (60–200 lb/ft²). The support posts are usually about 3–4 m (10–14 ft) apart in both directions.

Mezzanines impose heavy concentrated loads on the floor below. The load at each post is typically 3.5–12 t (8000–26 000 lb, or 4–13 US tons). The legs are far enough so that each load can be considered on its own, and for this reason a given floor can support a greater load at a mezzanine post than at a rack-leg. The area beneath the mezzanine will be used for some purpose, however, and this may complicate the structural design. It is usually possible – and always good practice – to avoid placing a mezzanine post near a slab edge.

Unlike pallet racks, mezzanine posts are typically equipped with substantial baseplates that have real load-spreading ability.

If the posts are very heavily loaded, it may be more practical to treat them as building columns with separate foundations. This approach is particularly useful where the other floor loads are light.

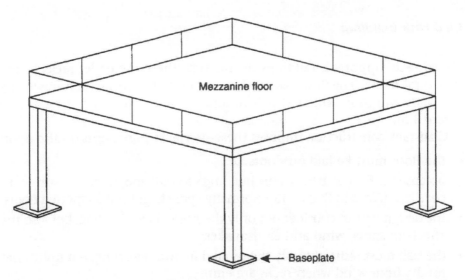

Figure 3.4 A mezzanine imposes widely separated loads on substantial baseplates

Here are the requirements for floors that support mezzanines:

- loading: concentrated loads, usually 3.5–12 t (8000–26 000 lb, or 4–13 tons) at each post;
- joints and cracks: no need for sealing, but posts should be located away from joints;
- surface regularity: not critical;
- wear resistance: not critical.

Mobile racks

These are pallet racks that slide horizontally on rails. Ordinary pallet racking requires at least one aisle for every pair of racks. In mobile racking many racks can be bunched together, with the possibility of opening an aisle at any point.

Here are some floor requirements for mobile racks:

- loading: the Concrete Society (2003) tells us to expect line loads of about 150 kN/m (10 000 lb/ft);
- joints and cracks: not critical;
- surface regularity: flatness is not critical, but levelness should normally be $F_L 20$ or better;
- wear resistance: not critical.

Clad-rack buildings

Called rack-supported buildings in America, these use pallet racks as the structural frame. The floor goes in first, followed by the racks, followed by the walls and roof, which attach to and are supported by the racks (see Figure 3.5).

Clad-rack construction imposes these special requirements on the floor:

- the floor must be laid outdoors;
- because the floor slab forms the building's foundation, it may be subject to code requirements that do not normally apply to ground-supported floors;
- the designer must consider not only the normal rack loading, but also the effects of snow, wind and earthquakes;
- the slab must resist not only the usual downward loading, but uplift that results from wind when racks are empty.

Figure 3.5 A clad-rack (also called rack-supported) warehouse under construction

Materials-handling systems

Most materials-handling systems use wheeled vehicles that ride directly on the floor. In some warehouses the vehicle loads control the floor's structural design. But even where they do not, vehicles impose requirements for joints and cracks, surface regularity and wear resistance.

Materials-handling equipment can be divided into these types:

- conveyors;
- low-level vehicles;
- counterbalanced forklift trucks;
- clamp trucks;
- reach trucks;
- turret trucks;
- order pickers;
- stacker cranes;
- hybrid vehicles;
- wire-guided vehicles.

Conveyors

These demand little of the floor. Goods are carried on rollers or belts supported by fixed steel frames. Because there are no moving parts in contact with the floor, conveyors have more in common with storage racks than with other materials-handling equipment.

Conveyors impose concentrated loads at fixed locations, much like the loads from storage racks. But the loads are usually lighter than storage-rack loads and rarely control the floor design.

Here are the floor requirements for conveyors:

- loading: variable, but usually less than for storage racks;
- joints and cracks: not critical;
- surface regularity: not critical;
- wear resistance: not critical.

Low-level vehicles

This category includes a wide range of vehicles including hand-pushed pallet jacks, motorized pallet transporters, trolleys and tractor-trailer combinations. Their common features are low height and relatively light loads.

Despite their light loads, some of these vehicles are surprisingly hard on floors. Pallet jacks and some motorized pallet transporters ride on very small, hard-plastic wheels that take a toll on floor joints and cracks. The wheels have to be small to fit under the pallet. Because pallet jacks ride with their forks close to the floor surface, they often bottom out on bumps. That problem is particularly acute with long-forked double pallet jacks that carry two pallets at the same time. Some trolleys and trailers ride on concrete-eating steel tyres, though that is not as common as it used to be.

Here are the floor requirements for pallet jacks and hand trolleys:

- loading: usually less than 500 kg (1100 lb) per wheel;
- joints and cracks: load transfer is important, and joints and cracks more than 1 mm (1/32 in) wide should be filled with semi-rigid epoxy or polyurea;
- surface regularity: F_F15/F_L13 for better, with special attention to curled-up joints where pallet jacks may scrape;
- wear resistance: AR2 or better (AR1 in areas of very heavy traffic).

Counterbalanced forklift trucks

These are the vehicles the ordinary person thinks of as forklifts. Load-bearing forks stick out in front. The forks move up and down, but do not rotate or extend relative to the rest of the truck. Because of that, counterbalanced trucks need wide aisles for maneuvering. They can serve either block stacking or pallet racks, but in either case the minimum aisle width is about 4 m (13 ft).

Most counterbalanced trucks ride on rubber or soft plastic tyres. These are harder than pneumatic tyres, but not quite so punishing to the floor as hard plastic or steel tyres.

Counterbalanced trucks are not very sensitive to surface irregularities. But problems can arise in high-speed operations, where loads sometimes bounce off the forks. A different problem can occur if trucks travel with the forks held very low. The forks may scrape high spots, damaging the floor surface.

Here are the floor requirements for counterbalanced trucks:

- loading: usually less than 2.5 t (550 lb) per wheel;
- joints and cracks: load transfer is important, and joints and cracks more than 1 mm (1/32 in) wide should be filled with semi-rigid epoxy or polyurea;
- surface regularity: $F_F 15/F_L 13$ or better, with higher numbers desirable for high-speed operation;
- wear resistance: AR2 or better (AR1 in areas of very heavy traffic).

Clamp trucks

These closely resemble counterbalanced forklifts, on which they are based, but they clamp loads from the sides instead of lifting them on forks. They are used for goods that come packed in big boxes – household appliances, for example.

The floor requirements for clamp trucks are the same as those for counterbalanced forklifts, except in the area of surface regularity, where clamp trucks demand less. Clamped loads will not bounce off no matter how rough the floor, and the trucks rarely bottom out at bumps.

Reach trucks

The reach truck has forks that extend forward relative to the rest of the truck. Compared to a counterbalanced forklift, the reach truck can use a narrower aisle and often stacks to a greater height. The typical aisle width is about 2.5 m (8 ft).

Reach trucks are sometimes called narrow-aisle trucks, but the term can cause confusion. To most people in the field of materials handling, a narrow-aisle vehicle is one that needs an aisle of about 2.5 m (8 ft) wide. Reach trucks meet that definition. Some people, however, use *narrow-aisle* for turret trucks and order pickers that need no steering room at all and can operate in an aisle less than 2 m (6 ft) wide. In this book, I adopt what I believe to be the more common practice. I use *narrow-aisle* for reach trucks, and *very-narrow-aisle* (often abbreviated VNA) for turret trucks and order pickers.

Reach trucks ride on hard plastic tyres, which easily damage open joints and cracks.

The need for surface regularity varies with lift height. Low-rise reach trucks do not need much more than counterbalanced forklifts, but a tall model is almost as demanding as a turret truck of similar height. Table 3.1 gives recommended tolerances.

Here are the floor requirements for reach trucks:

- loading: usually less than 2.5 t (5500 lb) per wheel;
- joints and cracks: load transfer is essential, and joints and cracks more than 1 mm (1/32 in) wide should be filled with semi-rigid epoxy or polyurea;
- surface regularity: see Table 3.1;
- wear resistance: AR2 or better (AR1 in areas of heavy traffic).

Turret trucks

These forklifts operate in an aisle only slightly wider than the truck itself. The forks are mounted on a turret that lets them rotate in the horizontal plane and reach out to each side. Some turret trucks stack to great heights. Lift heights of 9 m (30 ft) are common, and some models go to about 15 m (46 ft). With many designs, the operator's cab rises with the load.

Table 3.1 Floor surface regularity for reach trucks

Lift Height	Recommended Surface Regularity	
	F-numbers	TR 34
Up to 5.5 m (18 ft)	$F_F 20/F_L 15$	FM3
5.5–8 m (18–26 ft)	$F_F 30/F_L 20$	FM2
Over 8 m (26 ft)	$F_F 50/F_L 30$	FM1

Because the side clearance is so small – typically 100 mm (4 in), VNA trucks cannot be steered manually while in the stacking aisles. They are steered automatically, either by guide rails or by wire guidance. Outside the stacking aisles, VNA trucks are usually steered manually. A few systems are totally driverless and keep the trucks on guidance even when transferring from aisle to aisle.

Turret trucks demand a high degree of floor surface regularity. Many people assume this is to prevent the tall mast leaning over and striking the racks, but the situation is more complex than that.

Turret trucks have hard tyres and rigid, unsprung frames. That combination is needed to reduce mast sway – imagine the effect if a 9 m (30 ft) mast were supported on luxury-car springs. It makes the vehicles highly sensitive to floor irregularities, however. Turret trucks get a hard ride even on a fairly smooth floor.

On unflat floors, turret trucks wear out early and sometimes fail structurally. Problems can occur even where the floor is flat and level enough to keep the truck from striking the racks.

Table 3.2 shows recommended tolerances.

The need for surface regularity is most critical in the aisles, where turret trucks operate at full speed and full height, though not necessarily at the same time. The aisles should be treated as defined-traffic areas for the purposes of specifying and measuring surface regularity (see Chapter 19).

The need is less critical in the open areas, where the trucks usually go slower and keep the forks low. Except in fully automated systems, the open areas should be treated as random-traffic floors.

Table 3.2 lists recommended tolerances for the defined-traffic aisles where turret trucks run.

Here are the other floor requirements for turret trucks:

- loading: up to 5.6 t (12 300 lb or 6.2 US tons) per wheel (the heaviest loading occurs when a turret truck has extended its load into the racks);
- joints and cracks: load transfer is essential, and joints and cracks more than 1 mm (1/32 in) wide should be filled with semi-rigid epoxy or polyurea;
- wear resistance: AR2 or better (AR1 in areas of very heavy traffic).

Table 3.2 Floor surface regularity for defined-traffic warehouse vehicles (including turret trucks, order pickers and hybrid vehicles) – Concrete Society recommendations

Lift Height	Floor Classification
Up to 8 m (26 ft)	Category 2
8–13 m (26–40 ft)	Category 1
Over 13 m (40 ft)	Superflat

Order pickers

The lift trucks described so far normally carry palletized goods, or objects similar to pallets in size. Order pickers, in contrast, let workers fill orders by picking small objects from racks and shelves. An order picker may carry a pallet on which the operator can put the picked goods, but it does not place the pallet in the racks. For that reason, order pickers are rarely the only lift trucks in a warehouse. The typical warehouse using order pickers also contains counterbalanced forklifts, reach trucks or turret trucks to fill the racks.

Order pickers can work in aisles of almost any width, from the 4-m (13-ft) aisles used by counterbalanced forklifts to aisles even narrower than the 2-m (6-ft) aisles served by turret trucks. Order pickers in wider aisles are steered manually. Where the aisles are less than 2.5 m (8 ft) wide, the vehicles are steered by guide rails or wire guidance.

Order pickers seldom control the floor design, because other warehouse vehicles usually impose higher loads.

Stacker cranes

These are stacking vehicles that ride on rails. There is usually one rail at the bottom to bear the weight, and another at the top to keep the crane upright. Some models hang from overhead rails. Like a turret truck, a crane can operate in a very narrow aisle about 2 m (6 ft) wide. Cranes go higher than trucks; some stack to 25 m (80 ft).

Cranes are not easily shifted from aisle to aisle. Most users do not even try; they just buy one crane for every aisle.

Cranes are almost never manually operated; they normally form part of a computer-controlled automated storage-and-retrieval system (ASRS).

Because they run on rails, stacker cranes demand little of the floor. Here are the requirements:

- loading: usually not critical, because rack loads control the design;
- joints and cracks: not critical;
- surface regularity: not critical because rail is levelled independently of the floor (but some ASRS manufacturers impose overall levelness tolerances to simplify installation);
- wear resistance: not critical.

Hybrid stackers

These very-narrow-aisle vehicles combine some features of stacker cranes and turret trucks. Like cranes, they have tall non-telescoping masts that engage a top rail while in the aisles. Like turret trucks, they ride on the floor and can travel freely from aisle to aisle. Hybrids can stack to about 20 m (66 ft).

Some models ride on a steel channel when in the aisles, but it does not greatly change the way they interact with the floor. Unlike crane rails that are made of heavy steel and levelled independently of the floor, the channels used with hybrids are thin and follow the floor's general contour. Their purpose is to guide the vehicle, not to level or support it.

Hybrids are similar to the highest-stacking turret trucks in what they demand of a floor. Here are the requirements:

- loading: as high as several tons per wheel, but rack loads are almost always higher and control the floor design;
- joints and cracks: load transfer is essential, and joints and cracks more than 1 mm (1/32 in) wide should be filled with semi-rigid epoxy or polyurea – except for systems in which all vehicle wheels ride on a steel channel;
- surface regularity
  - in the stacking aisles: $F_{min}80$–100, or TR 34 Superflat Category;
  - in open areas: $F_F 50/F_L 30$;
- wear resistance: AR1 or better.

Wire-guided vehicles

This category overlaps some of the others. It consists of vehicles that follow an electric wire embedded in the floor. Almost any vehicle could be wire-guided, but in practice the technology is limited to turret trucks and order pickers in very-narrow-aisle (VNA) warehouses, to low-level pallet transporters in automated systems, and to some specialized factory vehicles.

Guide wires are installed after the concrete has hardened – sometimes long after. The installer saws a groove about 10 mm (3/8 in) deep, puts the wire in it and seals it with hard epoxy. Some systems use several wires together, requiring a deeper cut. The result looks like a sawcut joint but generally does not act as one.

Wire guidance affects floor design in three areas: embedded steel, movement at joints and surface regularity.

Steel lying too close to the guide wire interferes with the signal. To prevent that, wire-guidance suppliers often specify that steel reinforcement and post-tensioning tendons be kept a minimum distance from the wire, but they do not specify the same minimum. Recommendations have ranged from 50 mm (2 in) to an absolute prohibition on any embedded steel. That last requirement puts the floor designer in some rather tight handcuffs, as you can imagine.

In my experience, steel does not cause trouble unless it is very near – almost in contact with – the guide wire. (A big mass of steel farther away might create a problem, but few floors contain such a mass.) And reinforcing bars that cross the wire are far less likely to interfere than those that run parallel, or nearly parallel, to it. A design that keeps the steel 50 mm (2 in) from the wire (and remember that clearance can be horizontal as well as vertical) is almost certainly safe, if strictly followed.

Metallic dry shakes and steel fibres are usually allowed, on the assumption that the individual pieces of steel are small and evenly distributed around the guide wire. Similarly, few object to steel-armoured joints crossing the wire. However, some materials-handling consultants take a stricter view and would forbid all those things.

Slab movement at joints can break guide wires. The damaging movement may be either horizontal or vertical. Horizontal movement occurs at free contraction joints as the concrete undergoes drying shrinkage or thermal contraction. Vertical movement occurs as vehicles cross joints that lack good load transfer. If you know which joints are likely to move, you can tell the guide-wire installers, who can leave slack in the wire at those spots.

Wire guidance also affects the requirements for floor flatness and levelness. Floor irregularities can throw the vehicle off to one side. If this happens with enough force to overpower the automatic steering, the vehicle will leave wire guidance and may crash.

Because wire-guided vehicles travel in fixed paths, the floors they run on are treated as defined-traffic areas for the purposes of specifying and measuring surface regularity. That means they are specified by F_{min} in the F-number system, or by one of the "defined-movement" floor classes in the TR 34 system. F_{min} values for wire-guided vehicles generally range from 40 to 100, with the lower numbers specified for pallet transporters, and the higher for hybrid vehicles and the tall, fast turret trucks. In the TR 34 system, the Concrete Society (2003) directly ties the specification to lift height (see Table 3.2).

Here are the floor requirements for wire-guided vehicles:

- loading: varies according to vehicle type;
- joints and cracks: load transfer is essential and guide wire should be looped at joints where noticeable horizontal movement is seen or expected;
- surface regularity: at least $F_{min}40$ or TR 34 Category 2;
- wear resistance: varies with vehicle type.

Battery-charging areas

Many warehouse vehicles run on batteries, which have to be charged. The charging usually takes place in a dedicated area. People can and do charge batteries on ordinary concrete slabs, but attention to three details can improve the floor's performance in the battery-charging areas.

The first concern is with resistance to acid. Batteries contain sulphuric acid, which is highly corrosive on concrete. Over the years, some acid spills are bound to occur. To help resist them, floors in battery-charging areas are often covered with epoxy coatings or epoxy toppings.

The second detail is related to the first, and involves protecting the slab from subgrade moisture. Many warehouse floors are safely laid without damp-proof membranes to keep out subgrade moisture. But what works for the main warehouse floor may not be good enough for the battery-charging area. Moisture can make epoxy coatings and toppings peel off. If the battery-charging area will be coated or topped, the slab in that area should get a damp-proof membrane on all but the driest sites. And even if the original design does not include epoxy, it can be wise to install a damp-proof membrane if there is any chance the use may apply epoxy later.

The third detail only applies in those warehouse where the batteries are stacked on racks for charging. Some battery-charging systems employ a rail-guided lift truck to remove the batteries from the warehouse vehicles and stack them on the charging racks. Some of those trucks stack four levels high and need a very level surface. One manufacturer recommends a floor specification of $F_{min}85$ for its four-level battery stacker.

The load-class system

The best way to design a warehouse floor is to tailor it to the specific storage and materials-handling system that will be used. This not only ensures that the floor will serve its purpose, but often keeps the cost down as well. Some costly features, such as armoured joints or superflat tolerances, can be confined to those parts of the floor that really need them.

But not every floor can be tailored to a particular usage. Many warehouse users demand flexibility. And when a warehouse is built on speculation, the designer does not even know the user – much less the usage.

The situation is not without hope, however. Even when we do not know a floor's exact usage, we can often make reasonable assumptions based on the floor area and the building's clear height.

The load-class system, introduced by the UK's Building Research Establishment, standardizes those assumptions (Neal, 1987; Neal and Judge, 1987). We start out by assigning the floor a load class. There are four classes: light, medium, heavy and very heavy.

We then design the floor to support all the loadings specified for that load class. Tables 3.3 and 3.4 list the critical loads for each class. Once the floor has been certified as suitable for a particular load class, users present and future can freely change the usage without the need for further structural analysis, as long as they stay within the load limits for the specified class.

The latest edition of the popular UK guide *TR 34* (Concrete Society, 2003) backs away from the load-class system, saying: "It is strongly recommended

Table 3.3 The load-class system – SI units

Load Class	Critical Loads			
	Pallet Racks	Shelving	Mezzanine	Lift Truck
1. Light	4.5 t per end frame or 2.25 t per rack-leg	4.0 t per end frame or 2.0 t per leg	3.5 kN/m^2 on mezzanine	2.0 t capacity
2. Medium	6.0 t per end frame or 3.0 t per rack-leg	5.4 t per end frame or 2.7 t per leg	5.0 kN/m^2 on mezzanine	3.0 t capacity
3. Heavy	10.0 t per end frame or 5.0 t per rack-leg	Seldom critical	7.25 kN/m^2 on mezzanine	Seldom critical
4. Very heavy	12.0 t per end frame or 6.0 t per rack-leg	Seldom critical	9.5 kN/m^2 on mezzanine	Seldom critical

Table 3.4 The load-class system – US units

Load Class	Critical Loads			
	Pallet Racks	Shelving	Mezzanine	Lift Truck
1. Light	10 000 lb per end frame or 5000 lb per rack-leg	8800 lb per end frame or 4400 lb per leg	75 lb/ft² on mezzanine	4400 lb capacity
2. Medium	13 200 lb per end frame or 6600 lb per rack-leg	11 800 lb per end frame or 5900 lb per leg	100 lb/ft² on mezzanine	6600 lb capacity
3. Heavy	22 000 lb per end frame or 11 000 lb per rack-leg	Seldom critical	150 lb/ft² on mezzanine	Seldom critical
4. Very heavy	26 400 lb per end frame or 13 200 lb per rack-leg	Seldom critical	200 lb/ft² on mezzanine	Seldom critical

that the existing classification should be used with caution, particularly for more heavily loaded floors with combinations of high point loads from racking and MHE [materials-handling equipment]". I have no wish to enter that argument, but I will say this: whatever its flaws, the load-class system is the best tool we have for dealing with floor loading in speculative-warehouse construction. And it is miles ahead of practice in the USA, where developers have been known to specify 125-mm (5-in) slabs in warehouses with a 10 m (30 ft) clear height, and then act shocked when prospective tenants find the floors inadequate.

As useful as the load-class system can be, it does not cover – nor does it claim to cover – every aspect of floor design. It deals with loads and structural capacity, but does not address other important issues such as joints and cracks, surface regularity and wear resistance. Designers must still make judgments on these issues.

4

Factory and Special Floors

In Chapter 2 we looked at non-industrial floors, most of which are relatively easy to design and build because they are concealed by floorcoverings or applied floorings. Chapter 3 covered warehouse floors, which are often more demanding. In this chapter, we look at some of the most challenging floors of all: factory floors and floors for special purposes.

Factory floors

A factory is an industrial building used mainly for manufacturing or processing. Factories vary widely, ranging from small engineering workshops to food-processing plants to electronics assembly buildings. Because they vary so much, it is hard to generalize about their floors. We can talk about a "general-purpose" floor in a warehouse or office building, and we can even design and build such a floor on speculation, not knowing the user's identity. In contrast, the "general-purpose" factory floor does not exist. Every factory floor must be, or at least should be, designed to meet its user's specific needs.

Most factories include some storage space and are, to that extent, warehouses. They contain storage and materials-handling systems identical to those described in Chapter 3. But factories have their own special needs in the following areas:

- loading;
- joint sealing and filling;
- crack sealing;

- surface texture;
- wear resistance;
- resistance to chemical attack;
- dimensional stability;
- slope for drainage.

Loading

Some factory equipment is very heavy, imposing concentrated loads far greater than those found in warehouses. If the load exceeds a few tons, it is often best to provide a separate foundation for it. But in some cases it makes sense to put even very heavy equipment on the floor slab, particularly if the loads are many or if the user wishes to move the equipment from time to time.

Load safety factors for stationary loads on a factory floor can usually be set lower than those for moving vehicles such as forklift trucks. Floors designed for vehicle loads need a high safety factor to prevent fatigue failure, but fatigue is hardly an issue where a heavy machine stays in place for twenty years. Ringo and Anderson (1992, p. 9) suggest that safety factors can go as low as 1.3 for some stationary loads. The Concrete Society (2003, p. 54) recommends a partial safety factor (for use only with the design equations found in *TR 34*) of 1.2 for what it calls "permanent actions", which would include fixed factory equipment. However, higher safety factors may be needed where factory machines impart substantial vibrations to the floor slab.

Joint sealing and filling

We seal joints to keep them from trapping dirt or to stop liquids leaking through. We fill joints to protect the concrete edges from damage caused by traffic. Some factory floors need joint sealing or filling, or both, but many floors need neither.

The need for joint sealing varies from industry to industry. Companies that bash metal and saw wood rarely need sealed joints. Companies that process food and drugs usually do. In some cases government or industry regulations require sealed joints. In other cases users demand them because they make the facilities easier to keep clean and more likely to pass inspections.

Joint sealers should be highly elastic to accommodate movement, unless the need to withstand traffic dictates a harder material.

In contrast with joint sealing, the need for which depends on the products being made, the need for joint filling depends on the kind of traffic that crosses the joints. What matters most is not the overall loading but the contact pressure, which varies with the type of tyre. Pneumatic tyres have low contact pressures, equal to the air pressure to which they are pumped up. They do not require filled joints. Solid rubber tyres have higher contact pressures, and generally require filled joints unless the traffic frequency is low. Steel tyres – not so common today as in years past – impose the highest contact pressures and can destroy a joint in short order unless it is very well filled.

Only those joints exposed to wheeled-vehicle traffic need to be filled. Since we cannot always know the traffic patterns in advance, it often makes sense to wait till the factory is in operation before filling joints.

What do you do when the usage seems to demand both joint sealing and joint filling? That is a hard question, because no product on the market combines the elasticity of a sealant with the strength and hardness of a good filler. While awaiting that perfect product, consider these options:

- install a semi-rigid joint filler several months after construction, after most drying shrinkage has occurred;
- armour the joints to eliminate the need for a semi-rigid filler (but joints will still need a sealant);
- adopt a method of crack control that eliminates or greatly reduces the need for joints.

If you choose the last option, the possible methods include continuous reinforcement, post-tensioning, shrinkage-compensating concrete and a heavy dose of steel fibres.

Crack sealing

Factory users who demand sealed joints usually expect sealed cracks, too. Drug manufacturers and food processors sometimes insist that "all" floor cracks be sealed.

That seems clear, but it can lead to arguments because cracks come in all widths, down to the microscopic. To reduce disputes, specifications should state the critical crack width beyond which sealing is needed, and the time at which cracks are to be measured.

Surface texture

Most factory floors benefit from power-trowelling to a smooth, burnished finish. That produces an easy-to-clean surface with good wear resistance.

A burnished finish is wrong for some uses, however. It can become dangerously slippery if liquid or fine powder is spilled on it – a frequent occurrence in many factories. One of the most slippery floors I ever walked on was in a cigarette factory, where almost-invisible tobacco dust covered the surface.

Any of the following methods can be used to roughen the floor surface and improve slip resistance:

- give the floor a float finish;
- give the floor a broom finish;
- apply a slip-resistant dry shake;
- finish the floor by early-age grinding;
- roughen the finished floor surface by shot-blasting;
- score the finished floor surface with shallow saw cuts;
- apply a slip-resistant coating.

Users should know that all of these measures will make the floor harder to clean, and possibly less resistant to wear.

For more on floor finishing, see Chapter 17.

Wear resistance

Some factory floors need a high degree of wear resistance. While a floor in wear-resistance class AR1 is good enough for almost any warehouse, some factories require more.

The need is especially great in factories where abrasive materials are present on the floor surface. Metalworking shops are notorious for this, and the condition occurs in other factories as well.

To make floors wear-resistant, designers often use special hardeners, coatings and toppings. But plain concrete can wear surprisingly well, if well finished and well cured.

Though designers and builders bear the main burden of making factory floors wear-resistant, users too have responsibility. When an industrial process produces abrasive debris, an efficient floor-cleaning programme can pay rich dividends.

Resistance to chemical attack

Many industrial processes use chemicals that attack concrete. The list of troublesome chemicals includes some materials not commonly thought of as dangerous or corrosive. The food industry deals with many acids that pose not the slightest hazard to human health, but attack concrete severely. Without protection, a plain concrete floor in a meat-packing plant or a vegetable cannery will not last long.

Power-trowelling helps floors resist attack by densifying and strengthening the concrete's surface. But it only goes so far, because the vulnerability is rooted in concrete's basic chemistry. A more complete solution is to cover the concrete with another material such as epoxy or tile.

See Chapter 21 for more on chemical attack and how to resist it.

Special floors

Some floor uses impose special requirements. It is impossible to list them all, for a new use may be invented tomorrow. In the rest of this chapter, we shall look at floors for the following special situations:

- air-bearing vehicles;
- car parks;
- freezer buildings;
- television studios.

Air-bearing vehicles

These work on the principle of the hovercraft, supporting a load on a cushion of compressed air. Because a hovercraft covers a large area and uses low air pressure, it can travel over rough ground. In contrast, the air bearings used on floors cover a smaller area and use higher air pressure.

Air-bearing vehicles can be designed to work on almost any floor. But if the floor is level, smooth and airtight, then the air bearings can be smaller, the air demand lower and the overall system more efficient.

Floors that meet the following requirements have given good service in air-bearing systems:

- *joints and cracks:* all joints and obvious cracks should be sealed to reduce air leakage. Consider post-tensioning to eliminate cracks and keep joints in compression;

- *surface finish:* smooth trowel finish;
- *surface regularity:* $F_F 50/F_L 30$;
- *wear resistance:* not critical.

Multi-storey car parks (parking garages)

Some structures are built strictly for car parking, while others contain a mixture of parking levels and floors used for other purposes.

Our concern here is with the suspended floors. Many parking structures have a ground-supported floor on the lowest level (which is sometimes below grade), and that raises no special issues. (It may even be paved with asphalt.) But the upper floors demand special attention.

Car-park floors have two unusual requirements: the need for long spans, and the need to protect embedded steel from corrosion.

Parking structures benefit from very long spans to maximize open space. Closely-spaced columns annoy drivers and reduce the area available for parking. The need for long spans, coupled with the high loads imposed by cars that weigh 1–2 t (1–2 US tons) apiece, creates a challenge for the structural engineer. Some engineers use flat-plate floors, but often a one-way slab on beams is a better choice. Precast double tees are popular in some countries. See Chapter 6 for more information on these structural types.

Corrosion of reinforcing steel is a huge problem in car-park floors, especially in regions where the authorities salt roads to melt ice and snow. Cars drive in dripping with brine, which saturates the concrete, creating the perfect environment for corrosion of steel. The situation is even worse in a parking structure than on a road or bridge (though it can be bad there, too), because the indoor slabs are never washed clean by rain.

Designers fight such corrosion in different ways, with varying degrees of success. Some try to keep the salty water away from the concrete slab with waterproof coatings. Others try to make the concrete itself more water-resistant; adding microsilica to the mix is one way to do that.

Many American designers choose post-tensioning with unbonded tendons for car-park floors. Tucked away inside grease-filled plastic sheaths, the steel tendons are well protected from corrosion. The anchorages remain somewhat vulnerable, however, and demand careful attention. Epoxy-coated anchors are available, but they do not eliminate the need to seal the short exposed length of steel tendon where the sheath has to be stripped.

Factory and Special Floors 47

Freezer buildings

A freezer building is a warehouse kept at sub-freezing temperatures, usually for storing food. In the UK it is more often called a cold store, but I avoid that term here because it can also describe a chilled building kept slightly above the freezing point. Such a cold but unfrozen building poses no special challenges for the floor designer.

Because a freezer building is a warehouse, its floor is subject to all the normal requirements for a warehouse floor. As it is also a freezer, its designer needs to consider four extra points:

- frost heave;
- thermal contraction;
- difficulty of repairs;
- insulation.

Frost heave

In the early days of freezer construction, before people understood the risk, concrete slabs were cast right on the ground. The ground quickly froze, and any water present expanded as it turned to ice. As time passed, the frost went deeper and deeper, with more and more water turning to ice. Eventually the frozen subgrade formed a thick lens of icy soil heaving up the centre of the freezer floor. That was frost heave, and it destroyed many a slab.

Nowadays, freezer slabs are almost always laid over a thick bed of stiff foam insulation, but insulation alone only slows, and does not prevent, frost heave.

The most complete solution is install a sub-floor heating system, either in the sub-base or in a mud slab. (A mud slab is rough, thin concrete slab laid under the main floor slab.) Insulation boards separate the heated layer from the finish floor slab. The floor freezes, but the heated sub-floor does not.

Not all freezer buildings need sub-floor heaters. Some designers get by without them where the subgrade is dry and well drained. Eliminating the heater is not a decision to be made lightly, however, since the consequences of getting it wrong are so severe.

Thermal contraction

A freezer floor undergoes more thermal contraction than other floors, because it has to be laid at a temperature much higher than it will experience in operation. Suppose we lay a freezer slab 100 m (328 ft) long with the concrete at 20°C (68°F). If we chill the building to −20°C (−4°F) – and many freezers are even colder than that – thermal contraction alone will make that floor about 45 mm (1-3/8 in) shorter. That shortening has to be accommodated somewhere – at joints, at cracks or at the slab ends, depending on the floor design.

Ordinary floors also shorten over time, due mainly to drying shrinkage. But freezer floors shorten more because they are affected to thermal contraction in addition to the drying shrinkage that affects all concrete. For that reason, freezer floors benefit from a conservative approach to crack control. Some designers choose post-tensioning or heavy continuous reinforcement. Other designers rely on control joints but space them closely.

Because thermal contraction widens joints and cracks, they should not be filled till the building has been lowered to its final working temperature. That limits the choice of fillers, since some set very slowly when cold.

However, the news with regard to thermal contraction in freezers is not all bad. Unlike most other kinds of buildings, freezers experience almost no temperature changes after they go into use. For that reason, joint and crack fillers tend to work well and last a long time in freezer floors, provided they are installed after the slab has reached equilibrium.

Difficulty of repairs

Freezer floors are hard to repair, and that fact makes good design and good workmanship especially important.

In an ordinary warehouse or factory, it sometimes makes sense to build a cheap floor and accept the need for repairs and higher maintenance costs. If joints spall, they can be patched. If the surface dusts, it can be sealed.

Applied to a freezer floor, that sort of thinking leads to disaster. Many chemicals used to patch or seal concrete do not harden or cure well at low temperatures. Even purely mechanical repairs such as grinding and sawing are hard to carry out inside a freezer, especially if they rely on water. Though some repairs are possible, a freezer floor – more than almost any other kind – simply must be laid right the first time.

Insulation

Modern freezer floors are almost always laid over thermal insulation. It usually takes the form of rigid boards of closed-cell plastic foam, laid in two courses with the seams staggered.

Figure 4.1 Freezer floors usually include a thick course of insulation under the slab

Like the insulation used in walls and roofs, the boards used under floor slabs are specified by their insulating value (see Figure 4.1). In the UK, insulating value is typically expressed as thermal conductivity, stated in watts per metre kelvin. In this system lower values mean better insulation, and the materials used under freezer floors measure about 30 W/(mK). Americans look at insulation the other way around, specifying resistance to heat transfer, called R-value. R-value is measured in $(h)(ft^2)(°F)/BTU$. As you might guess, everyone just calls it R-value. When the same property is measured in SI units, it is called R_{si}. To convert R to R_{si}, multiply by 0.176. In this system higher values mean better insulation, and commonly available products have an R-value of about 5 per inch (R_{si}, 0.035 per mm). Insulation is usually installed at least 100 mm (4 in) thick, for an R-value of 20 (R_{si}, 3.5). Some designs call for thicker insulation. In most cases, the engineer who designs the freezer's refrigeration system will specify the amount of insulation required under the floor.

The material used under the slab must be more than just a good insulator, however. It must also be waterproof and strong enough to support the concrete slab and loaded racks. Some closed-cell plastic foams can meet both those requirements. They come in various formulations with compressive strengths ranging from 200 to 700 KPa (30 to 100 psi).

When deciding what compressive strength to specify, it is not good enough simply to match the strength to the loading. We have to apply a generous safety factor. Dow Chemical, a manufacturer of freezer-floor insulation, recommends safety factors of 3 for static loads, and 5 for dynamic loads. Tall, clad-rack freezer buildings impose loads that approach the capacity of the strongest available insulation boards. Indeed, the strength of the insulation may turn out to be the limiting factor in the height of such buildings.

Television studios

Television and video recording studio floors have two special requirements: flatness and sound-proofing. In addition, some studio users like to cover their floors with epoxy, imposing a few extra requirements to prevent problems with that material.

Flatness

Studio floors need flatness, so cameras can roll smoothly without bumps and ripples that affect image quality. Levelness matters much less than flatness (see Chapter 19 for the distinction between flatness and levelness). Unfortunately, studio users do not agree on how much flatness they really need.

Some just ask for the best they can get, if not more. Back in the days when the flattest floors were routinely, if wrongly, specified at 3 mm (1/8 in) under a 3-m (10-ft) straightedge, some television people arbitrarily demanded a tolerance twice as strict – 1.5 mm (1/16 in) under the same 3-m (10-ft) straightedge. Few if any studio floors ever met that standard, but people continued to specify it. Today those demanding users will more likely pick the highest specification of any list provided them, or to ask whether something even better is available.

At the other extreme, some studio users make do with ordinary floors and wonder what all the fuss is about.

In the face of such widely varying demands, we can hardly talk about any standard requirement for floor flatness in television studios. But a few points seem clear.

First, uniformity of flatness is essential. Users vary in the degree of flatness they require, but almost all will complain if one part of the floor – usually at a joint – is markedly worse than the rest. Since floors are almost always less flat near construction joints, it is a good idea to design studio floors without such joints. If the overall flatness requirements are so high that the floor must be divided into narrow strips – likely where the flatness specification exceeds F_F70 – grinding will probably be needed near the joints.

Second, some techniques developed for superflat warehouse floors do not work on high-tolerance floors for studios. The reason is that superflat floors in warehouses, almost without exception, support defined-traffic aisles. The floor is cast in strips, with a superflat aisle running down the centerline of each strip. Since the superflat tolerances apply only in the aisle, concrete finishers have developed techniques aimed at creating flat aisles, sometimes at the cost of flatness elsewhere. The typical superflat warehouse floor looks very flat if measured down the aisles, but less flat – often far less – if measured at right angles to the aisles. That would never do in a television studio, where the users insist on equal flatness in all directions. To make a superflat floor flat in all directions for studio use, it may be necessary to make repeated passes with a hand straightedge riding on the forms, adding mortar to offset the effect of plastic settlement.

Third, any joint or crack more than about 2.0 mm (0.08 in) wide needs to be filled to make it effectively invisible to the video cameras as they cross it. Unlike a warehouse floor where the filler serves to protect the joint from damage, the main goal here is to make the joint as smooth and flush as possible. Ordinary industrial joint fillers – semi-rigid epoxies and semi-rigid polyureas – work well in studios, but the standard instructions are not always adequate to ensure a perfectly smooth and flush result. The surest method is to overfill the joint or crack, let the filler harden, and then grind or sand it down flush with the concrete surface.

Sound-proofing

This may or may not be important, depending on where the studio is located. The strictest requirements for sound-proofing exist where a single building contains more than one studio – a fairly common situation.

The usual way to sound-proof a floor is to build it as a floating screed laid over acoustic insulation. The insulation should also be used at the edges of the topping.

Part II

Structural Design

5

Structural Design of Ground-Supported Floors

Concrete floors are divided into two main structural types. Ground-supported floors rest on the ground and are supported by it. Suspended floors are supported by other building elements.

Not all floors at ground level are ground-supported. Some are actually suspended floors, supported by piles. This chapter does not apply to them.

This chapter deals with:

- components;
- basic principles;
- structural types;
- design methods;
- design considerations.

Components

Figure 5.1 shows the components of a ground-supported floor. Some are optional, but every ground-supported floor includes subgrade and slab.

The subgrade is the underlying ground that supports the floor and all floor loads. It may be undisturbed natural soil or imported fill. The quality of the subgrade is an important factor in the structural design of a ground-supported floor.

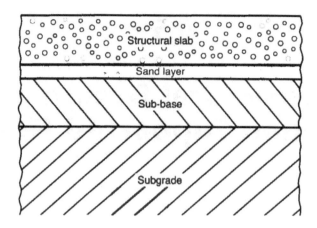

Figure 5.1 The components of a ground-supported floor

Some subgrades supports slabs directly, but in many cases a sub-base is laid over the subgrade. Designers specify sub-bases for these reasons:

- to increase the floor's load-carrying capacity, by spreading concentrated loads over a larger area of the subgrade;
- to provide a stable base for construction traffic and erection of forms;
- to create a free-draining layer beneath the slab.

Sub-bases most often consist of compacted granular material, typically crushed rock with a broad mix of particle sizes. Lean concrete, soil cement and other materials are also used.

In some floors, a layer of fines covers the sub-base. Ordinary sand is often used, but crushed-rock fines maintain grade better under construction traffic (see Figure 5.2). The fines smooth irregularities in the sub-base, making the slab thickness more precise and reducing friction between slab and sub-base. The fines layer typically has a nominal thickness of 50 mm (2 in) or 75 mm (3 in), but its actual thickness varies greatly because of bumps and dips in the sub-base. The fines layer may be redundant where the sub-base contains enough fine material of its own for smooth grading (see Figure 5.3).

Some designers use fines for another purpose – as a blotter over a damp-proof membrane, to soak up excess water from the concrete mix. When a damp-proof membrane is used directly under a floor slab, it can lead to excessive bleeding and increased curling. Putting a layer of dry fines between slab and damp-proof membrane reduces the risks. On outdoor pours, however, a blotter layer may cause more grief than it prevents. A heavy rainfall will soak the blotter layer, which then takes a long time to dry.

Structural Design of Ground-Supported Floors 57

Figure 5.2 All-in crushed rock, with a blend of coarse and fine particles, makes a good sub-base

Figure 5.3 A coarse-aggregate sub-base provides good structural support and drains well, but is hard to grade precisely

Plastic sheeting is sometimes placed under the concrete slab for two separate purposes: as a slipsheet to reduce friction between slab and sub-base, and as a damp-proof membrane to prevent transmission of water vapour from the ground into the slab. When used as a slipsheet, the plastic always goes directly beneath the slab. When used as a damp-proof membrane, it may be placed either directly beneath the slab, or between the fines layer and the sub-base, or even under the sub-base. The use of plastic sheeting is controversial, and is discussed in more detail in Chapters 16 and 22.

The slab, made of concrete, is the floor's main structural component and the one that gets most of the attention. It transmits loads to the sub-base (if used) and ultimately to the subgrade, spreading them out over a big enough area that deflections are small. In many buildings, the slab also provides the wearing surface.

Toppings (see Chapter 18) and floorcoverings (see Chapter 22) may lie above the slab.

Structural types

Ground-supported floors can be divided into five structural types:
- unreinforced slabs;
- slabs reinforced only for crack control;
- structurally reinforced slabs;
- post-tensioned slabs;
- slabs made of shrinkage-compensating concrete.

Unreinforced slabs

These contain no steel reinforcement except dowels at joints. Such floors are rare in the UK but common in America. They rely wholly on the concrete's strength to support loads. The essential principle is that the load-induced stress within the slab must not exceed the concrete's flexural strength (modulus of rupture). If properly designed, these slabs do not crack under load. They may crack for other reasons, but not from loading. Engineers who work exclusively in reinforced-concrete may find the concept strange, but it has a long history and is very widely used for concrete roads.

Slabs reinforced only for crack control

These contain steel reinforcement, but are designed as if they did not. The reinforcement – steel bars, wire mesh or steel fibres – helps control shrinkage cracks but does not add, or is assumed not to add, to the slab's structural capacity. Most floors in this category are only lightly reinforced, but exceptions exist. For example, jointless floors made with continuous reinforcement may contain as much as 0.5% reinforcing steel, by cross-sectional area, in two directions. Nevertheless, most designers still treat them as unreinforced when analyzing their structural strength.

This category includes slabs made with shrinkage-compensating concrete (though some authorities put them in a class of their own). The small amount of prestress created by the cement's expansion, and the modest amount of reinforcing steel normally added to such slabs, do not add significantly to the floor's load capacity.

Traditionally, most ground-supported floors in the UK and America were reinforced only for crack control, but that may be changing. The UK has seen a trend, encouraged by the popular third edition of *TR 34* (Concrete Society, 2003), toward steel-fibre designs in which the fibres are taken into account in determining slab thickness. The USA has seen a contrary trend toward totally unreinforced slabs.

Structurally reinforced slabs

These contain enough reinforcing steel that the designer takes it into account when analyzing the slab's load-bearing capacity. The reinforcement can take the form of steel bars, wire mesh or steel fibres.

The analysis usually follows standard principles for reinforced-concrete design, that means that unlike the slabs in the first two categories, structurally reinforced slabs may crack under load. The designer aims not to prevent load-related cracks, but to ensure the floor has enough reinforcement to resist the stresses after such cracks have occurred.

Floors for very heavy loads, as in clad-rack buildings, are often designed with top and bottom layers of steel reinforcement. Floors for lighter loads are sometimes designed with just one layer, normally near the bottom.

Slabs reinforced with steel fibres fall in this category if the fibre dosage is high enough to make a real difference in the slab's post-crack performance, and if the designer chooses a design method that takes that difference into account.

Post-tensioned slabs

Post-tensioned slabs are prestressed with steel tendons. The *pre* in prestress comes from the fact the slab is deliberately stressed *before* working loads are applied. The *post* in post-tensioning comes from the fact that the tendons are stretched *after* the concrete has set. Post-tensioning puts slabs in compression, producing an effect similar to a rise in the concrete's flexural strength. The slab is usually designed as if it were unreinforced, but with allowance for the higher equivalent flexural strength, taken to be the sum of the concrete's inherent flexural strength (modulus of rupture) and the residual prestress midway between slab ends.

Post-tensioning allows some very efficient designs, with less steel than an equivalent structurally-reinforced slab, and less concrete than an equivalent unreinforced slab. Its popularity has remained strictly regional, however. Though widely used in Australia and the American Gulf Coast, it is hardly even considered in broad areas of the world.

Design methods

Design methods for ground-supported floors fall into three types:
- finite-element analysis;
- standard methods found in design manuals;
- designs based on experience.

Finite-element analysis

This involves breaking down the loads and floor components into tiny elements, which are then analyzed using basic principles of structural engineering. It used to be both difficult and expensive. While it still demands a high degree of engineering skill, modern computers have made it much more accessible.

Finite-element analysis would seem to promise the most elegant, most efficient way to design a concrete floor. It has not, however, proven consistently superior to simpler, rougher methods, and there are two reasons for that. One reason is that the structural behavior of a ground-supported slab is very complex and not fully understood. Engineers employing finite-element analysis have to make simplifying assumptions, the same

Structural Design of Ground-Supported Floors 61

as in the traditional methods. The other reason is that many of the variables in floor design are poorly controlled and impossible to predict with any precision.

Standard methods

While a few designers use finite-element analysis, far more choose a standard method found in a design manual. There are many to choose from. Because you cannot follow a standard method without reading the design manual that presents it, I do not explain any of the methods in detail here. I confine myself to general comments that may help you decide which method to use.

UK standard methods

In the UK, choices include:
- Chandler's *Design of Floors on Ground* (1982), which should be accompanied by Chandler and Neal's *The Design of Ground-supported Concrete Industrial Floor Slabs* (1988);
- the *TR 34* method, set out in the Concrete Society's Technical Report 34, *Concrete Industrial Ground Floors* (2003).

Chandler and Neal offer a simple method explained with great clarity. (Chandler, 1982; Chandler and Neal, 1988). It gives us the choice between working out answers from equations or looking them up in tables. The method leans heavily on old work by Westergaard (1925, 1947), Kelley (1939) and Pickett, most of which involved pavement design for vehicle loads. Despite that, it contains an up-to-date treatment of rack and shelving loads for warehouses.

Chandler and Neal's method works well for unreinforced slabs, and for slabs that contain light reinforcement or light dosages of steel fibre for crack control. It can also be used for slabs with heavy reinforcement or heavy dosages of steel fibres, but may result in their overdesign, since it does not account for any structural contribution from the steel.

If someone told me I could only use one design method, this is the one I would pick.

TR 34 (Concrete Society, 2003) provides a more complex method based on plastic analysis, similar in some ways to that applied to suspended floors in draft Eurocode 2. In contrast, most other standard methods, including Chandler and Neal's, rely on elastic analysis. The use of plastic analysis presumes that the concrete slab will stay ductile after cracking, and that in turn requires a substantial amount of steel fibres, steel reinforcement near the slab's bottom or prestress. *TR 34* implies that the normal ground-supported slab will contain steel fibres, though an Appendix extends the method to slabs reinforced with wire mesh. You cannot use *TR 34* to design unreinforced slabs. Nor can you use it for slabs that contain reinforcement only near the top, even though that is the position many authorities recommend for crack control.

The *TR 34* method is a good choice if you are designing an industrial floor with steel fibres and wish to take advantage of their effect on concrete's post-crack ductility. That advice holds even if you work in America, since the standard American methods do not account for the structural benefit of steel fibres at high dosages. The method is of limited use beyond that.

American standard methods

Standard design methods (see Table 5.1) in America include:

- the PCA (Portland Cement Association) method, set out in *Concrete Floors on Ground* (Spears and Panarece, 1990);
- the Corps of Engineers method, set out in *Engineering Design: Rigid Pavements for Roads Streets, Walks and Open Areas* (Corps of Engineers, 1984);

Table 5.1 Standard design methods

Method	Suitable for Unreinforced Slabs	Slabs Reinforced for Crack Control	Structurally-reinforced Slabs	Post-tensioned Slabs
Chandler and Neal	X	X		
Corps of Engineers	X	X		
PCA	X	X		
PTI				X
Ringo and Anderson	X	X	X	X
TR 34			X	
WRI	X	X		

Structural Design of Ground-Supported Floors

- the WRI (Wire Reinforcement Institute) method, set out in *Design Procedures for Industrial Slabs* (Wire Reinforcement Institute, 1973);
- the PTI (Post-Tensioning Institute) method, set out in *Design and Construction of Post-Tensioned Slabs on Ground* (Post-Tensioning Institute, 1980);
- Ringo and Anderson's *Designing Floor Slabs on Grade* (1992).

The PCA method is widely used for warehouse floors. In principle, it resembles Chandler and Neal's method, but the two differ greatly in presentation. The PCA method relies heavily on complex charts rather than tables and equations. Different charts deal with vehicle loading (single and dual axles), rack loading and block loading. One limitation of this method is its presumption that loading occurs in the slab's interior – that is, away from edges and corners. If you can't avoid edge loading, this method gives you little choice but to thicken the slab there.

The PCA method works for unreinforced slabs and for slabs reinforced for crack control.

The Corps of Engineers method offers a good way to design slabs for vehicle loads. It considers edge loading and allows for the effect of load transfer at joints. It works for unreinforced slabs and for slabs reinforced for crack control.

The WRI method resembles the PCA method. Though it assumes slabs will contain crack-control reinforcement (hardly a surprise, since the publisher is the Wire Reinforcement Institute), it works for unreinforced slabs, too.

The PTI method deals with post-tensioned floors. It presents a good treatment of the stresses created by swelling soil, and how to resist them with post-tensioning. Ringo and Anderson (1992) point out that while the PTI manual presents post-tensioning as the solution to swelling soil, you can use it (the manual) even if you choose to resist the stresses with steel reinforcement instead of post-tensioning. The PTI method does not offer a way to calculate the slab stresses created by normal floor loading. For that, you need one of the other design manuals. If designing a post-tensioned warehouse floor, say, you would normally use another manual to determine the load-induced stresses, and the PTI method to determine the post-tensioning needed to resist those stresses.

Ringo and Anderson's book (1992) is a design manual with a difference. It summarizes and refers to all the American design methods listed above, as well as drawing on some other less familiar sources. Ringo and Anderson present several approaches but also identify their favourites, calling them "Authors' choice". The result is a book that belongs on every floor designer's shelf, and not just in America. Non-American readers should be warned, however, that Ringo and Anderson concede nothing to the metric system; everything here is in US measurements. If you work in SI units, get out your conversion tables.

The Australian option

For those willing to stretch beyond the UK and America, the Cement and Concrete Association of Australia publishes the useful *Industrial Floors and Pavements*. It offers a clear explanation and the choice between simple and rigorous methods. The simple method is simple indeed: a single table in which to look up slab thickness, meant for "lightly loaded industrial pavements". The rigorous method provides design equations and graphs for concentrated loading (from vehicles and racks) and block loading with aisles. The latest Australian thinking holds that traditional design methods for block loading sometimes resulted in too-thin slabs.

Using a standard design method

To a degree, each standard design method requires its own approach, spelled out (though not always as clearly as you might wish) in its manual. Even so, some generalizations are possible. The design process typically includes these steps:

1. Collect information on the loads. The main items of interest are the magnitude of each load, the area of loading and the distance between loads.
2. Determine the type of load. The usual choices are concentrated loads (as from rack legs or forklift wheels), block loads with aisles (as from stacked pallets or rolls of paper), uniformly-distributed loads and line loads (as from walls). Each type calls for a different analysis.
3. Collect information on the subgrade. The main item of interest is the modulus of subgrade reaction.
4. Decide whether or not to modify the subgrade.
5. Assume a concrete strength. Flexural strength is the key property, but some design methods use compressive strength as a proxy for it.
6. Assume a slab thickness.
7. Determine the tensile stress induced in the slab by the design load plus a safety factor. If appropriate, determine stress at internal, edge and corner positions on the slab, with allowance for load transfer at joints.

8. Check for the effect of nearby loads, which usually add to the tensile stress, but sometimes reduce it.
9. Compare the stress induced by loading to the concrete strength.
10. If the stress and strength match, the design is sound.
11. If the stress is much less than the modulus of rupture, the floor is overdesigned. Go back to Step 6 and try a thinner slab.
12. If the stress is greater than the modulus rupture, the floor is underdesigned. Go back to Step 6 and try a thicker slab. In some cases, it may make sense to change the assumptions made in Steps 3 and 4.

Designs based on experience

After reading the last few pages, you might think that most floor designs were based on a careful analysis of slab stresses. In fact relatively few floors are so designed. For the majority of ground-supported floors around the world, structural design is based not on calculations but on experience – either the designer's own experience, or the collective experience recorded in building codes and national standards.

Most ground-supported floors in houses, offices and shops are designed by minimum-thickness rules. Many light-duty and medium-duty industrial floors are also designed by these rules.

BS 8204:Part 2 recommends a minimum slab thickness of 100 mm (4 in), no matter how light the loads. ACI 302 provides a bit more information, recommending different minimum thicknesses for various floor uses. See Table 5.2.

A 150 mm (6 in) slab thickness has become, in many places, the standard thickness for industrial floors where the loadings are modest or unknown. It would not have maintained its popularity if it didn't work, most of the time. Whether by coincidence or the result of trial and error, a 150 mm (6 in) slab laid on reasonably good subgrade turns out to be just about right for the average forklift bearing an average load.

Table 5.2 Minimum thickness for ground-supported floor slabs (American Concrete Institute)

Floor Use	Minimum Thickness
Houses, offices, institutions	100 mm (4 in)
Light industrial and commercial	130 mm (5 in)
Industrial	150 mm (6 in)

My old boss, Allen Face, used to joke that the way to design a floor was to follow this equation:

Slab thickness = Westergaard, Kelley, Pickett and all the other standard formulae $\times 0 + 6$ in

That is not so far from the way thousands of successful floors have been designed. It *is* a joke, of course, and no one seriously suggests that a 150 mm (6 in) slab can work under all conditions.

But even where heavy loads dictate thicker slabs, some designers rely more on experience than on calculations. A few owners specify slab thickness in heavily loaded warehouses, based on what worked for them on previous projects. If a firm has a dozen similar warehouses, all with, say 200 mm (8 in) slabs, it's hardly unreasonable to conclude that the thirteenth warehouse should get the same design.

Design considerations

In the rest of the chapter, we shall look at some of the considerations, major and minor, that affect the structural design of ground-supported floors.

Types of loading

When a floor is designed for specific loads, one of the early steps is to identify the type of loading. There are at least five types:

- nominal loading;
- uniformly-distributed loading;
- block loading with aisles;
- concentrated loading;
- line loading;
- horizontal loading.

Nominal loading is an arbitrary figure used for design purposes. It has little to do with actual, measured floor loads. It is usually expressed in terms of a uniformly distributed load, in kN/m^2 or lb/ft^2 – even though the actual loads may be anything but uniformly distributed.

Nominal loadings appear often in building codes and users' requirements. They can be useful in residential, office and institutional buildings, where the specific loads are unknown but almost always light.

Uniformly-distributed loading exerts even force over a large area. The purest form of this kind of loading comes from liquids in tanks and from loose bulk materials such as corn and gravel. The category can also include certain floors on which pallets or boxes are stacked. But if the goods are stacked in blocks, with aisles left open for access, the loading falls in the next category. Uniformly-distributed loads are usually easy to accommodate because they create little bending stress within the slab. They are expressed in kN/m^2 or lb/ft^2.

Block loading with aisles exists where goods are stacked on the floor with unloaded areas (the aisles) left open for access. It is common in warehouses that lack racks and shelves. The critical stress normally occurs at the aisle centerline, and depends to some extent on the aisle width. Block loading is expressed in kN/m^2 or lb/ft^2.

Concentrated loading, also called point loading, exists where a weight bears on an area of less than $1\,m^2$ ($10\,ft^2$). Loads from wheeled vehicles always fall in this category. Loads from racks and shelving usually belong here, but in a few cases are treated as line loads instead. All concentrated loads are analyzed similarly, except that wheel loads often get higher safety factors than do loads from racks and other fixed equipment (see Figure 5.4).

In the UK, concentrated loads are usually expressed in tons (t), and occasionally in kilonewtons (kN). In America, they are expressed in US tons, pounds (lb) or kips. (1 kip = 1000 lb)

Line loading is concentrated loading stretched out along a line. The most common source is a wall resting on the floor slab. Line loading also occurs with some rack systems and stacker-crane rails. Most rack systems involve concentrated loads, however. Where line loads are substantial and far apart, providing separate footings for them may prove more efficient than beefing up the floor slab. Line loads are expressed in kN/m or lb/ft.

All the loads discussed so far are vertical. Floors also withstand *horizontal loading* from vehicles that are braking, accelerating or turning. Vehicles that travel freely on the floor exert horizontal loads through their tyres, and the loads are then limited by the friction between tyre and floor. Since the coefficient of friction is almost always less than 1, horizontal loads under those conditions do not exceed the vertical loads. Vehicles that ride on or are guided by rails exert horizontal loads through the rails, and those loads could be higher than loads from free-ranging vehicles. Nevertheless, most designers ignore horizontal loading, and there seems to be little risk in doing so. If horizontal loading at guide rails creates trouble, failure is far more likely at the connection between rail and slab than within the slab itself.

Figure 5.4 Pallet racks impose concentrated loads

Load safety factors

It would not be wise to design a floor so that it barely supported its load without cracking. We need a margin of safety to allow for errors in design or construction, and, most of all, to prevent fatigue failure. The safety factor is always greater than 1 (if it were less, the slab would be designed to fail). We can apply the safety factor to the loading, in which case we multiply the load by safety factor, or we can apply the safety factor to the concrete strength, in which case we divide the strength by the safety factor. Some design tables have safety factors built in, while other do not.

Structural Design of Ground-Supported Floors

If we are not careful, we can apply the safety factor more than once, ending up with an overdesigned floor.

Choosing the safety factor is one of the most important decision a floor designer faces. It has a huge effect on the final design and on the floor's cost, and it lies almost wholly under the designer's control. Rarely do owners or builders review or even know the load safety factor on which the floor design is based.

Recommended safety factors range from 1.2 to about 3.0. Low numbers are used for fixed loads where all important variables are well controlled. High numbers are used for repetitive loading and uncertain conditions. Chandler (1982, p. 7) recommends at least 1.5, even for fixed loads. Ringo and Anderson (1992, p. 9) recommend 1.4 to 2.0 (see Table 5.3), but say: "When in doubt, use 2.0".

Some authorities closely link the safety factor to the predicted number of load cycles, to prevent fatigue failure (see Table 5.4). A factor of 2.0 is supposed to suffice for unlimited load cycles.

But just what constitutes a load cycle? The answer is less clear than you might think. A traffic engineer would probably take it to be one vehicle crossing the floor one time. Chandler (1982), no doubt visualizing warehouse trucks, says it involves a vehicle travelling laden in one direction, and returning unladen in the opposite direction. That seems clear enough for forklift loads, but what about a warehouse where rack loads control the design? Rack loads fluctuate, but only rarely, if ever, make a complete swing from totally empty to fully packed. If anyone has developed a good method for counting load cycles in rack loading, I have yet to hear of it.

When evaluating load safety factors, designers should remember that "safety" means different things in ground-level and elevated (upper-storey) construction. When elevated floors fail, buildings collapse and people die. Because the consequences of underdesign are so severe, safety factors for elevated slabs are not left to the designer, but are imposed by building codes.

Table 5.3 Load safety factors for ground-supported slabs (adapted from Ringo and Anderson, 1992)

Safety Factor	Application
1.4	Non-critical and low-traffic areas with no impact loading and few load cycles
1.7	Important areas with many load cycles, where design variables are under good control
2.0	Critical areas with many load cycles, where one or more design variables are not well controlled

Table 5.4 Load safety factors to prevent fatigue failure (Chandler, 1982)

Safety Factor	Maximum Number of Load Cycles Over Floor's Lifespan
2.00	Over 400 000
1.96	400 000
1.92	300 000
1.87	240 000
1.85	180 000
1.82	130 000
1.79	100 000
1.75	75 000
1.72	57 000
1.70	42 000
1.67	32 000
1.64	24 000
1.61	18 000
1.59	14 000
1.56	11 000
1.54	8000
1.50	Under 8000

Ground-supported floors, in contrast, cannot fall down and almost never (I'm tempted to say never, but someone will surely find an exception) fail catastrophically. At ground level, structural failure means the slab cracks under load and may settle to an objectionable degree. Such failure is undesirable and we should try to prevent it, but it is clearly far less serious than structural failure in suspended, elevated floors.

Please do not take my comments as a license to ignore safety factors. We need them, and too high is always better than too low. But low safety factors – low compared to those that apply to suspended floors – can often be justified for ground-supported floors.

Area of loading

This matters in the analysis of concentrated loads. If a given weight is spread over a larger area of floor surface, it creates less tensile stress in the slab.

Some design calculations use the area of loading in mm^2 (in^2). Other methods use the radius of the area of loading, which is the radius of a circle equivalent in area to the area of loading. Still others use what's called

Structural Design of Ground-Supported Floors 71

the equivalent radius of the area of loading, which takes slab thickness into account and differs, usually by a small amount, from the actual radius. It is easy to become confused about all this, because the commonly used design methods are not consistent. Some methods use the actual area or radius and make the necessary adjustments for the designer. Other methods use the effective radius and expect the designer to calculate it before using the equation of entering the chart.

Where the load comes from a single post on a thick, rectangular baseplate, finding the area of loading is as simple as measuring the baseplate. But some situations are more complex.

In the case of wheel loading, the contact patch is hard to measure directly. If the wheel has a pneumatic tyre, the area of loading will be, almost exactly, the air pressure divided by the load. If the tyre is solid rubber or plastic – common on warehouse vehicles – the area will be smaller than for an equally loaded pneumatic tyre, but the exact size depends on the composition and thickness of the tyre material. The vehicle manufacturer may be able to provide details.

Where two concentrated loads lie close together – separated by not more than twice the slab's thickness – we can assume they both bear on a single area that includes the area between them. This condition occurs with back-to-back pallet racks and dual wheels on industrial trucks. This substantially increases the area of loading for design purposes.

Rack and shelving legs normally rest on baseplates, and that fact creates both an opportunity and a risk with regard to area of loading. The opportunity is the option to specify a bigger baseplate, reducing stress within the slab. The risk is that the steel plate making up the baseplate may be too thin to do much good. To spread load, the baseplate must be stiffer than the slab, but the thin sheet steel often used does not meet that requirement. Truth be told, most baseplates are designed not to spread load but to create a place for the anchor bolt. Unless you know the baseplate is stout, the safe approach is to ignore the baseplate and assume that the area of loading does not stretch beyond the bottom of the post.

Load position

Concentrated loads have different effects in different positions on the slab. A load located well away from the slab edge – so-called internal loading – creates the least stress in the slab. The same load will produce more stress if located near a slab edge, and will produce yet more if located at a corner slab.

The differences are substantial. Suppose we apply a 1 t (2200 lb) load to a floor slab 150 mm (6 in) thick. In this example the concrete has a flexural strength of 3.8 MPa (550 psi) and the sub-base is good. The radius of area of loading is 100 mm (4 in). According to Chandler (1982), that load in an internal position produces flexural stress of 0.70 MPa (102 psi). If we move that load to the slab's edge, the stress rises to 1.10 MPa (160 psi) – a 57% increase. If we move it to the corner, the stress reaches 1.28 MPa (186 psi) – 83% higher than for internal loading.

We can deal with the edge-and-corner problem in four ways:

- design the whole floor for the worst condition, which is usually corner loading;
- avoid loading the slab edges and corners (hard to do where vehicle loads control the design);
- strengthen the slab edges – usually by thickening them, but sometimes by adding reinforcement;
- transfer load to adjacent slabs through dowels, tie bars or reinforcement.

Effect of nearby loads

Some concentrated loads – those from mezzanine posts, for example – are far enough apart that each one can be considered on its own. More often, however, loads occur within a metre or two of each other and must be considered in combination.

If the distance between two loads, centre to centre, is no more than twice the slab thickness, we can treat them both as one, including the space between them as part of the loaded area.

If the loads are farther apart, normal practice is to determine the stress resulting from more demanding of the two, and then adjust that figure for the effect of the second load. If both loads are identical, we start with the one in the more critical position. Edge loading is always more critical than internal loading, and corner loading is usually more critical than edge.

With edge and internal loading, adjacent loads always increase stress, and the amount of increase diminishes with distance. But where the critical load occurs at a corner, adjacent loads can either raise or lower the stress, depending on their distance from the critical load, on the slab's stiffness, and on the subgrade support.

The standard design methods provide ways to cater for adjacent loads, though in some cases the treatment is more implied than explicit. Chandler (1982) is unusually clear on this subject.

Structural Design of Ground-Supported Floors

Subgrade properties

Because the subgrade ultimately supports all floor loads, its properties have a large effect on the structural design of the floor. We need to consider:

- the modulus of subgrade reaction;
- settlement;
- expansion and contraction from moisture changes.

Modulus of subgrade reaction

This is a measure of the ground's ability to resist immediate elastic deformation under load. It is defined as the pressure needed to deform the subgrade by a given unit distance. In the UK, the modulus is often reported in meganewtons per cubic metre (MN/m^3) or newtons per cubic millimetre (N/mm^3). But the correct SI expression is megapascals per metre (MPa/m), and that is the unit used here. Americans use pounds per cubic inch (pci).

You can measure the modulus of subgrade reaction directly by making a plate-bearing test on site. But because no great precision is needed, many designers just estimate the value from a general description of the subgrade material. Table 5.5 shows typical results for various soil types.

Some soils engineers measure the California bearing ratio (CBR) instead of the modulus of subgrade reaction. CBR values are widely used in road-building. Table 5.5 shows the relationship between CBR and modulus of subgrade reaction.

A higher value for subgrade modulus may allow a thinner slab, but the overall design is not very sensitive to small changes in the modulus.

Table 5.5 Modulus of subgrade reaction and CBR for typical soils

Subgrade	Description of Soil	Modulus of Subgrade Reaction	CBR
Excellent	Gravels and sandy gravels	82 MPa/m (300 pci)	25–80%
Good	Sands, gravelly sands, silty sands and clayey sands	54 MPa/m (200 pci)	10–40%
Poor	Very fine sands, silts and clays with plasticity index under 50	27 MPa/m (100 pci)	4–15%
Very poor	Silts and clays with plasticity index over 50	14 MPa/m (50 pci)	3–5%

If the modulus of subgrade reaction is too low, the following steps can raise it:

- remove the subgrade and replace it with better material;
- stabilize the subgrade by adding cement or lime;
- add a sub-base (see Table 5.6);
- use vibro-compaction.

Where the modulus is extremely low, the only reasonable solution may be to install piles and erect a suspended slab on them.

Adding a sub-base has the same effect as raising the modulus of subgrade reaction, but only for concentrated loads (see Table 5.6). A sub-base does not help support uniformly distributed loads or block loading with aisles.

Settlement

The modulus of subgrade reaction describes the soil's short-term, elastic response to loading. That is the subgrade property we need to know to determine slab thickness in most design methods.

A subgrade under load also experiences long-term, plastic deformation called settlement. Slab movement caused by settlement is often many times as great as the short-term elastic deformation predicted by the modulus of subgrade reaction.

By its nature, long-term settlement is hard to predict from tests. Even geotechnical engineers with local experience can predict it only within broad limits.

Settlement is not hard to live if it is uniform over the whole floor. Some floors have settled by more than 40 mm ($1^1/_2$ in) without causing any problems. The troublemaker is differential settlement, either within the floor or between the floor and other parts of the building. Differential settlement can cause wide cracks and lead to faulting at joints and cracks. Good isolation joints help prevent or minimize damage caused by differential settlement between the floor and other building elements. Differential settlement within the floor slab is a harder nut to crack. Heavy reinforcement or post-tensioning will reduce the distress, but may not be a complete solution.

Though we can never eliminate settlement, these measures can reduce its ill effects:

- choose a site where little settlement is expected;
- remove the subgrade and replace it with better material;
- pre-load the subgrade for several months before laying the floor.

Table 5.6 Effect of sub-bases on modulus of subgrade reaction

Sub-base Material	Sub-base Thickness	Effective Modulus of Subgrade Reaction where Original Modulus of Subgrade Reaction is:					
		14 MPa/m (50 pci)	20 MPa/m (75 pci)	27 MPa/m (100 pci)	40 MPa/m (150 pci)	54 MPa/m (200 pci)	60 MPa/m (225 pci)
Granular	150 mm (6 in)	18 MPa/m (66 pci)	26 MPa/m (96 pci)	34 MPa/m (125 pci)	49 MPa/m (181 pci)	61 MPa/m (225 pci)	66 MPa/m (243 pci)
	200 mm (8 in)	22 MPa/m (81 pci)	30 MPa/m (111 pci)	38 MPa/m (140 pci)	55 MPa/m (203 pci)	66 MPa/m (243 pci)	72 MPa/m (265 pci)
	250 mm (10 in)	26 MPa/m (96 pci)	34 MPa/m (125 pci)	44 MPa/m (162 pci)	61 MPa/m (225 pci)	73 MPa/m (269 pci)	81 MPa/m (298 pci)
	300 mm (12 in)	30 MPa/m (111 pci)	38 MPa/m (140 pci)	49 MPa/m (181 pci)	66 MPa/m (243 pci)	82 MPa/m (302 pci)	90 MPa/m (332 pci)
Cement-Bound	100 mm (4 in)	35 MPa/m (129 pci)	60 MPa/m (221 pci)	75 MPa/m (276 pci)	100 MPa/m (368 pci)	—	—
	150 mm (6 in)	50 MPa/m (184 pci)	80 MPa/m (295 pci)	110 MPa/m (405 pci)	—	—	—
	200 mm (8 in)	70 MPa/m (256 pci)	105 MPa/m (387 pci)	—	—	—	—
	250 mm (10 in)	90 MPa/m (33w2 pci)	—	—	—	—	—

Effective values apply only for concentrated loads – not for uniformly distributed loads.

Expansion and contraction from moisture changes

This is a serious problem with certain subgrade materials known as expansive clays. Many soils expand slightly when wet and shrink on drying, but expansive clays show this property in extreme form. In bad cases, the ground surface moves vertically by as much as 1m (3ft). The problem is most acute in climates with marked dry and wet seasons.

No simple solution exists, but all of the following have proven successful in some cases, either alone or in combination:

- post-tensioning;
- reducing moisture changes;
- replacing the expansive soil;
- supporting the slab on piles and carton forms.

Two-way post-tensioning does not stop the movement, but helps the slab keep its integrity as it moves up and down. Post-tensioned house slabs have become popular in parts of the southern USA where expansive clays are widespread. The Post-Tensioning Institute publishes a guide, *Design and Construction of Post-Tensioned Slabs on Ground* (1980) with much information on expansive soil and the use of post-tensioning to deal with it.

Another approach, popular in Australia, is to minimize moisture changes under the floor. Wide roof overhangs help keep rain away from the building's foundation, while careful attention to landscape planting reduces drying caused by roots and wetting caused by irrigation. The Cement and Concrete Association of Australia (2003), in its recommendations for houses built on expansive soil, suggests keeping all plants at least 1.2 m (4 ft) from the floor slab and planting trees. Lightly watered lawns area allowed beyond 1.2 m (4 ft). Trees and shrubs should be planted no closer to the house than 1.5 times their mature height (see Figure 5.5).

Removing the expansive clay is another possible solution – and a highly effective one, if you go deep enough. But in many cases huge amounts of material must be replaced. More than once I have seen a geotechnical engineer recommend replacing large amounts of soil, only to have the owner reject the recommendation after the bids came in.

An excellent solution, where money is plentiful, is to abandon ground-supported construction and design instead a suspended floor on piles. Provided the piles go below the expansive layer, you can design this sort of floor just like any other slab on piles. You cannot build it like any other, however. Most pile-supported slabs are cast directly on the ground, even though they do not rely on the ground for structural support. That will

Structural Design of Ground-Supported Floors

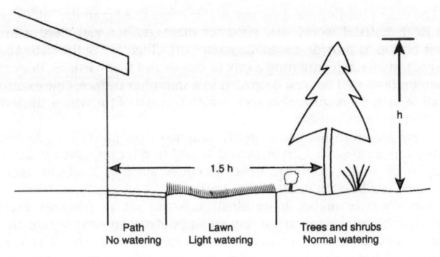

Figure 5.5 One way to deal with expansive clay is to control the landscaping adjacent to the floor slab

not work on expansive clay, because swelling soil would lift the slab. The answer is to pour concrete on corrugated paper forms, sometimes called carton forms. When dry, the paper forms are strong enough to support the concrete's weight, but after they become damp they disintegrate quickly.

Sub-bases

Many floor designs include a sub-base between the subgrade and the slab. The sub-base can serve at least three purposes.

First, it reduces slab stresses by spreading concentrated loads over a larger area of the subgrade. In effect, the sub-base raises the modulus of subgrade reaction (see Table 5.6). But this only works under concentrated loads; the sub-base does not improve the subgrade's ability to support uniformly distributed or block loads.

Second, a sub-base can provide a stable platform for construction traffic and for the erection of forms. This purpose often justifies a sub-base even where the working loads will be light.

Third, a sub-base can provide a free-draining layer beneath the slab. This helps keep the slab dry, and sometimes substitutes for a damp-proof membrane. But you need to be careful about that.

The most common sub-base material is compacted granular fill. Coarse granular material works well good for drainage, but may need a fines layer on top to provide a smooth surface on which to lay the slab. All-in granular material, containing a mix of coarse and fine particles, does not drain quite so well but can be graded to a smoother surface. One example of all-in material is *crusher-run* rock, which consists of the whole, unsieved product of a rock crusher.

A few American designers specify rounded river gravel for sub-bases. This offers unsurpassed drainage and is said to offer improved structural support because it conforms better to curled slabs. It is all but useless, however, as a base for construction traffic.

Lean concrete makes an excellent sub-base for all purposes except drainage. It contains much less cement and water than everyday concrete. The cement content is typically about 100 kg/m^3 (170 lb/yd^3). The mix is compacted and levelled with a vibrating roller. Lean concrete costs more than other sub-base materials, but often allows an offsetting saving in the thickness of the structural slab (where concentrated loads control the design).

Concrete strength

Chapter 7 covers this subject in more detail. But it is worth noting here that the load capacity of a ground-supported floor depends on the concrete's flexural strength, not its more commonly measured compressive strength. Most designers do not specify flexural strength directly; they specify compressive strength instead and rely on an assumed relationship between the two kinds of concrete strength.

We can specify flexural strength directly, however, and there are advantages to doing so.

Anything that raises concrete's flexural strength will improve a floor's structural capacity, even if compressive strength is unaffected or goes down. More work is needed, but it appears that certain coarse aggregates have this effect. Hayward (1985) reports good results from irregularly shaped aggregates such as crushed limestone.

Steel fibres improve flexural strength, but only at very high dosages. When added in more usual amounts – at 45 kg/m^3 (75 lb/ft^3) or less – they have a negligible effect on strength as we normally measure it. However, some design methods take advantage of the way steel fibres improve so-called ductility – defined as concrete's resistance to bending after it has cracked.

Structural Design of Ground-Supported Floors

Slab thickness

On most jobs, specifying the slab thickness is the most important structural decision. Of all the variables that affect a floor's structural capacity, thickness has the most powerful effect. If we double the subgrade modulus, the structural capacity barely rises. If we double concrete strength, the capacity also doubles (more or less). But if we double the thickness, we increase the capacity by a factor of about four.

Thicker slabs support more, but they also cost more. If calculations lead to a slab thickness that seems excessive, consider the following changes to reduce thickness and, possibly, make a more economical floor:

- improve the subgrade;
- thicken the sub-base (only effective for concentrated loads);
- strengthen base – switching to lean concrete, for example (again, only effective for concentrated loads);
- use a stronger concrete;
- use post-tensioning to increase the concrete's effective flexural strength;
- use bigger baseplates under concentrated loads.

Controlling slab thickness

Specifying slab thickness can be hard, but making sure the slab meets the specification can be even harder. Designers worry most about slabs that are too thin, but excessive thickness causes problems, too. A thick slab wastes concrete and may crack in unforeseen places.

Some designers specify tolerances for the slab thickness. In America, ACI 117 imposes a thickness of tolerance of +10 mm, −6 mm (+3/8 in, −1/4 in) for slabs up to 300 mm (12 in) thick. That means that a nominal 150 mm (6 in) slab could be as thick as 160 mm (6–3/8 in) or as thin as 144 mm (5–3/4 in). The tolerances are slightly looser for slabs more than 300 mm (12 in) thick. In the UK, BS 5606 allows a similar range, suggesting a thickness tolerance of +15 mm (+5/8 in).

It is questionable whether those tolerances have any effect other than to reassure designers and users. When slabs are measured (most are not), almost all are found to have sections that are too thin and too thick.

In real life, slab thickness is controlled not by the thickness tolerances, which floorlayers widely ignore, but by the elevation tolerances on the top and bottom of the slab. In other words, slab thickness depends on the accuracy of the grading and the elevation of the finished floor surface. Both properties vary greatly.

Ordinary grading is less accurate than many designers realize. According to one study (John Kelly Lasers, 1989), a typical graded sub-base, ready for slab placement, showed a standard deviation of 14.9 mm (0.59 in) from design grade. If we assume that the elevations followed a statistically normal distribution, then we can estimate that more than 10% of that sub-base's area was more than 25 mm (1 in) above or below design grade.

Floor surface elevation is also less accurate than many suppose. Many ordinary and quite satisfactory floor surfaces vary by as much as 20 mm (3/4 in) from design grade.

If a dip in the slab surface 20 mm (3/4 in) low happens to fall above a bump in the sub-base 25 mm (1 in) high, the slab there will be 45 mm (1–3/4 in) thinner than the design value. That violates all standard thickness tolerances. If detected, it would lead to complaints and disputes, and maybe even replacement of the slab. And yet it would not be the result of any horrible error. It would just be the result of normal construction practice, and could happen to almost anyone.

Faced with so much variation in slab thickness, what can we do? There are three possibilities.

One approach is to allow for the variations by adding a generous safety margin to the required slab thickness. But we might need a safety margin as great as 50 mm (2 in) to ensure that almost all the slab meets the desired minimum thickness. In other words, if calculations tell us we need a 200 mm (8 in), we might have to specify 250 mm (10 in). Many would object to what appears to be an absurd overspecification, compared to past practice. But if we use a smaller safety factor or none, we are, in effect, betting on better-than-normal workmanship.

Another approach is to reduce the variations. The increasing use of laser-controlled machines, both for grading earth and for striking-off concrete, promises tighter control over both the bottom and the top of the floor slab. The makers of automatic, laser-guided graders claim they can meet a tolerance of ±6 mm (±1/4 in). Laser screeds offer the potential, if not always the reality, of similar control over the finished floor surface. Narrow-strip construction allows good control of slab thickness, but is almost never specified just for that purpose, since it costs so much.

The third approach, common in the UK and also used in America, is to specify a +0 tolerance on the sub-base or subgrade. In other words, the bottom of the slab is not allowed to go higher than its design elevation, though it can go lower by some (possibly large) amount. The Concrete Society (2003) recommends a grading tolerance of +0 mm, −25 mm (+0 in, −1 in). If enforced (and it often is not), a +0 tolerance biases the variations

Structural Design of Ground-Supported Floors 81

in thickness. It reduces the risk a slab will be too thin, but raises the likelihood a slab will be too thick.

Building loads

Many authorities recommend that ground-supported floors should be structurally isolated from all other building elements. That is good advice, because a floor so built is easier to analyze and behaves more predictably. But it is not always practical.

By necessity or choice, designers sometimes tie the floor to other building elements and rely on it to resists stresses from them. Examples include:

- portal-frame construction;
- raft foundations;
- retaining walls.

In every case, designers need to think about how the stresses combine, and should be wary of using standard design details that were developed for isolated slabs.

Portal-frame construction

In most buildings, the upright supports, whether columns or bearing walls, exert almost all their force in the downward direction. Portal frame buildings, in contrast, create a substantial horizontal force as the frames try to spread outward (see Figure 5.6). Designers often use the floor slab as a tension tie to resist the spreading forces. That puts the floor in tension, reducing its ability to support other loads and complicated crack control.

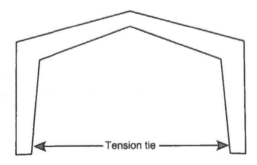

Figure 5.6 Portal frame

Raft foundations

Also called mat foundations, these are ground-supported floors that support the whole building. In most big buildings, each column rests on its own footing, which is isolated from the ground-supported floor. But if the column loads are heavy enough, or the soil weak enough, it can become economic to combine all the footings into one giant raft. The raft foundation usually does double duty as the ground-level floor.

Raft foundations can be designed along the same lines as other ground-supported slabs, with a few differences. Unlike ordinary floors, rafts are normally subject to all the code requirements for foundations. Those requirements may include rules for minimum reinforcement and distance between ground and reinforcing steel. Wind, snow and seismic loads need to be allowed for. In some cases (clad-rack buildings, for example), wind can create uplift, a condition almost never encountered in ordinary floors.

Retaining walls

Designers often rely on the floor slab to resist horizontal stress from retaining walls. That practice simplifies the design of the wall and complicates the design of the floor.

Warehouse floors are notorious for cracking in unpredictable ways around dock leveller pits. One factor (not the only one) in that cracking is the use of the floor to restrain the pit walls. Some designers reduce the problem by eliminating any structural connection between slab and walls, but they then have to find another way to restrain the walls.

6

Structural Design of Suspended Floors

A suspended floor rests on other building elements. Most suspended floors are on the upper storeys of multi-level building, and are supported by columns, beams or walls. Other suspended floors are at ground level, supported by piles. But most floors at ground level are ground-supported, not suspended.

Suspended floors and ground-supported floors are usually designed according to fundamentally different structural principles. As we saw in Chapter 5, most design methods used for ground-supported floors rely on the concrete itself to resist bending. Reinforcement is optional, and serves to control cracks rather than to increase structural capacity. Suspended floors, in contrast, always include structural reinforcement or prestress. The concrete is used to resist compressive stress (and some shear stress) only; tensile stress is resisted by the reinforcing or prestressing steel.

Unlike most ground-supported floors, suspended floors are an integral part of the building structure. For that reason, they are covered by the structural codes: Eurocode 2 in the UK and ACI 318 in America. Because a truly rigorous analysis of suspended-floor behavior is difficult, most designers use standard design methods taken from the codes. But the codes allow non-standard designs derived from basic principles, for those engineers who are so inclined.

This chapter is no more than a brief introduction to the structural design of suspended floors. A full treatment would demand a lengthy book of its own, and would require a writer more qualified than me. What I try to do here is present the basic ideas for readers who are not themselves structural engineers.

Types of suspended floors

Compared to ground-supported floors, suspended floors vary widely.

The most basic division is between one-way and two-way floors. A one-way floor transfers load to its supports in one direction only. A two-way floor transfers load in two perpendicular directions. According to the structural codes, any slab with a high enough aspect ratio (length divided by width) must be treated as one-way. But even a square panel is one-way if it contains structural reinforcement or prestress in one direction only.

Two-way floors can be built with less material, but one-way floors offer two advantages. Their length perpendicular to the span is unlimited. And in residential and light commercial construction where floors normally rest on walls rather than columns, one-way floors require fewer bearing walls. The non-bearing walls can be lighter and cheaper, and are more easily provided with openings.

Going beyond that basic division, we can consider the following structural types:

- one-way slabs;
- one-way joist floors;
- one-way precast floors;
- two-way beamless floors (subdivided into flat plates and flat slabs);
- two-way floors on beams;
- slabs on metal deck;
- other composite floors;
- ground-level floors on piles.

One-way slabs

In a one-way concrete slab, each suspended span (normally rectangular in plan) is effectively supported on two opposing sides by beams or walls.

Figure 6.1 shows a one-way slab. The slab spans in one direction between beams – which could just as easily be bearing walls.

The simplest one-way floor, common in residential construction consists of a slab supported on each end by bearing walls. A slightly more complicated system puts the slab on beams which are supported in turn by columns or walls. In beam–girder systems, the slab transfers its load to beams which in turn transfer the load to girders perpendicular to the beams.

The primary reinforcement or prestress in a one-way slab runs only on the direction of the span. Most one-way slabs contain secondary

Structural Design of Suspended Floors   85

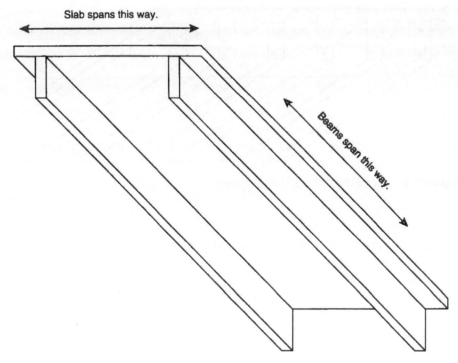

Figure 6.1 One-way suspended slab on beams

reinforcement in the perpendicular direction. The secondary reinforcement helps control cracks from drying shrinkage or thermal contraction. Because of its role in controlling thermal-contraction cracks, it is often called temperature reinforcement or temperature steel.

Some one-way slabs – particularly in slab–beam–girder systems – get structural reinforcement in the secondary direction to resist bending in that direction. But the amount of reinforcement is always less than in the span direction, and it is not considered to add to the load-carrying capacity of the slab.

One-way slabs can be designed efficiently for spans of up to 6 m (20 ft) between beams. There is no limit to the length of the slab in the other direction, at right angles to the span.

One-way joist floors

If the beams supporting a one-way slab are very close together – less than about 3 m (10 ft) apart – they are known as joists. Joists can be separate, but are often cast monolithically with the floor slab (see Figure 6.2). Slabs with

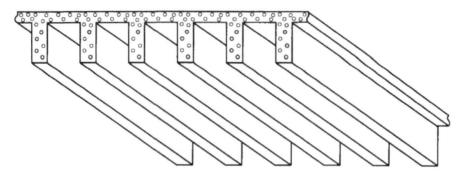

Figure 6.2 Suspended slab with integral joists

monolithic joists are sometimes known as pan-joist floors, because they are cast on steel forms called pans.

The strength of a joist system is determined mainly by the depth and spacing of the joists. Joists are typically 150–500 mm (6–20 in) deep and about 600 mm (2 ft) apart. To keep costs down, it makes sense to base the design on ready-made forming systems wherever possible, and that limits dimensions.

Joist floors can span up to 14 m (45 ft) in the direction of the joists. They can be designed for heavy loads and relatively small deflections, making them a good choice for many industrial applications.

One-way precast floors

Most concrete floors are cast in place, but the designer of a one-way floor can choose precast concrete instead.

To understand precast floors, it helps to consider how structural engineers analyze one-way floors. When designing a cast-in-place one-way floor, the engineer imagines it as a series of parallel beams 1 m wide (American engineers assume a beam width of 1 ft or 1 in). With a precast system we take this concept one step further and actually make the floor out of separate beams.

Precasting offers several advantages. The precast units can be mass-produced, reducing costs and possibly providing better quality control. Because the units are made in a factory, manufacturers can use techniques, such as pre-tensioning and steam-curing, that are seldom practical on building sites.

Some precast units – tees and double tees – incorporate the finished floor surface (see Figure 6.3). They are only suitable where numerous wide joints are not objectionable, which rules out many applications. Double tees are widely used in multi-storey car parks.

Other precast units – including those known as planks – are meant to receive a cast-in-place concrete topping. Figure 6.4 shows one version, but there are many others. Hollowcore planks are prestressed members with longitudinal holes to reduce weight. They typically come in widths of 1.2 and 2.4 m (4 or 8 ft) and thicknesses of 150 to 400 mm (6 to 16 in). Spans can reach 18 m (55 ft), but are usually far less.

The topping over planks is typically about 60 mm ($2^1/_2$ in) thick. It may contain light reinforcement for crack control. But if the topping will be left exposed – as is often the case in industrial buildings – expect to see a crack in the topping over every seam between planks, regardless of reinforcement.

The precast industry makes a wide range of tees and planks for suspended floors. Where possible, designs should be based on standard units. But special units can be ordered, and may be economic on some big jobs.

Figure 6.3 Section of precast double tee

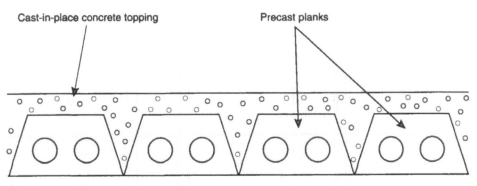

Figure 6.4 Precast concrete planks

Two-way beamless floors

These floors rest directly on the building columns (or on piles or pile caps, in the case of ground-level suspended floors). They are divided into two sub-types: flat plates and flat slabs.

Flat plates

The flat plate is the simplest two-way floor, being of constant thickness throughout (see Figure 6.5). Its main advantage is ease of construction. Formwork is easy to erect because the soffit (the floor's underside) is flat. Since formwork can account for more than half the cost of a floor, a design that allows simple forms has broad appeal.

Another advantage is that flat plates minimize storey height. Because there are no beams, girders or thickened sections under a flat plate, the overall floor thickness is less than with any other type.

Drawbacks include:

- inefficient use of materials, since the whole floor must be as thick as its most highly stressed part;
- large and unpredictable deflections.

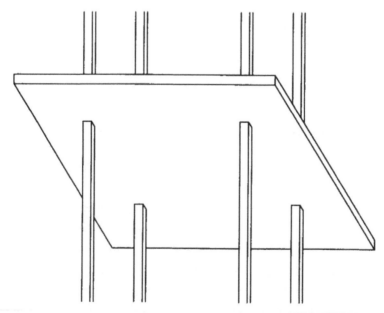

Figure 6.5 A two-way slab – flat plate

Australians Taylor and Heinman (1977) report on flat-plate floors that deflected by several times the predicted amount. Flat plates are not a good choice for floors that must be very level.

In general, flat plates make sense for lightly loaded floors with clear spans of up to 9 m (30 ft) in both directions. They are widely used for residential, office and institutional floors.

Flat plates are usually cast in place, but in *lift-slab* construction they are precast. Lift slabs are laid at ground level and then jacked up to their final elevated positions. Most lift slabs are flat plates.

Flat slabs

The flat slab differs from the flat plate in having thickened sections overthe columns or piles. The thickened sections help resist punching shear. The flat slab is harder to build, but uses materials more efficiently than the flat plate.

Compared to similar flat plates, flat slabs can support heavier loads over wider spans. They are sometimes used in factories and warehouses where the loads are too heavy for an efficient flat-plate design. For very heavy loads, however, most designers would switch to a one-way or two-way floor on beams.

Like flat plates, flat slabs are subject to deflections that are high and hard to predict. Where the user needs a very level surface, any beamless floor is likely to disappoint.

Flat slabs can be practical for spans of up to 10 m (34 ft) in both directions.

Two-way floors on beams

In these designs, each suspended panel is supported on all four sides by beams or walls.

Figure 6.6 shows a basic two-way floor with beams on every column line. The supports could just as easily be bearing walls. Some designers use intermediate beams in addition to the beams on the column lines. If the spacing between the intermediate beams is less than about 3 m (10 ft), the result is a waffle slab – similar to a one-way joist floor but with the joists running in two directions.

Two-way floors on beams generally deflect less than other suspended floors over similar spans. Waffle floors have a particularly good record

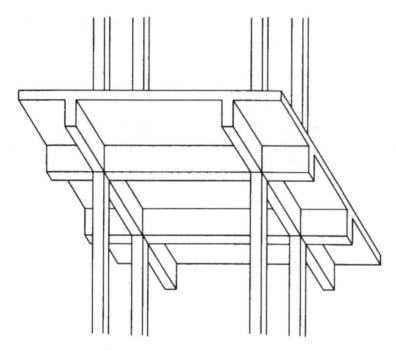

Figure 6.6 A two-way slab on beams

for stiffness. This makes them a good choice for users who demand a high degree of floor levelness. But no suspended floor is likely to be as level as a ground-supported floor built with equal care.

Two-way floors on beams can be designed for heavy loads and long spans. Waffle slabs can economically span up to 15 m (50 ft) in both directions.

The main drawback to these floors is that they are complicated to build. To minimize the difficulty and cost, they should be designed to use ready-made forms wherever possible.

Slabs on metal deck

These are concrete slabs laid over corrugated sheet steel. They are designed as composite floors in which both the steel deck and the concrete slab contribute to structural strength. They are considered one-way floors with the span running parallel to the corrugations.

Metal deck is also known as profiled metal deck, permanent steel formwork or – somewhat derisively – wrinkled tin. Normally zinc-plated, sometimes painted, it comes in various thicknesses ranging from 0.6 to 1.0 mm (0.024 to 0.040 in). The deck serves as a soffit form that need not be removed, and it reinforces the underside of the slab, eliminating the bottom mat of reinforcing steel used in other kinds of suspended floors. Metal deck is generally left unshored, but props may be needed where the span exceeds 3 m (10 ft). Many designs put the deck on closely spaced, steel-truss joists, keeping the spans short and eliminating all need for props.

Concrete thickness may be specified from either the top or the bottom of the deck corrugations, so it pays to ask. Thickness depends on load and span, but the Cement and Concrete Association of Australia (2003) recommends at least 125 mm (5 in) and suggests that thicknesses above 250 mm (10 in) are unlikely to be economic. Recent American practice favours thinner slabs – 100 mm (4 in) or even less. Very thin slabs add little to the floor's strength, and can be considered a sort of topping or screed to cover the steel. Thin slabs save money, but the results do not satisfy all clients. Some of these floors have experienced bad cracking and curling.

Slabs on metal deck typically contain light wire-mesh reinforcement. Heavily loaded floors are reinforced with deformed steel bars or heavy mesh. A few designs rely on steel fibres or no reinforcement at all (except the deck).

As they contain structural reinforcement or prestressing steel, ordinary suspended floors almost never contain the crack-control joints common in ground-supported slabs. But metal-deck floors are an exception. Though the practice is controversial, some people saw joints in these floors. Where used, control joints are normally put over the structural supports and at regular intervals in between, following the joint-spacing rules that apply in ground-supported construction.

Metal-deck slabs are cheap and easy to build, but they have some serious limitations. They are notorious for high deflections. The American standard for concrete tolerances, ACI 117, prohibits the enforcement of levelness tolerances on suspended slabs that are cast without props – a category that includes most metal-deck floors. Unless heavily reinforced (few are), slabs on deck are vulnerable to wide cracks and objectionable curling.

Other composite floors

The slab on metal deck is the most common type of composite floor, but not the only one.

Both one-way and two-way slabs on steel beams can be designed as composite floors. It is not good enough, however, for the slab simply to

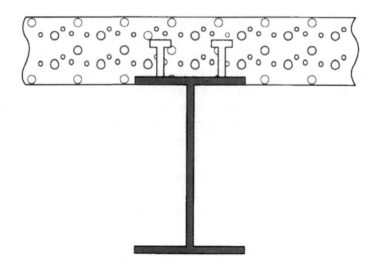

Figure 6.7 In composite construction, shear studs tie the slab structurally to the steel framing

rest on the beam. The two elements must be tightly connected, usually through shear studs (see Figure 6.7).

Ground-level floors on piles

These do not really form a separate structural type. They are usually designed as flat plates or flat slabs, but one-way or two-slabs on beams are also possible.

What distinguishes pile-supported floors from other suspended floors is their location at ground level. They are normally cast right on the ground, like ground-supported floors. They do not need soffit forms, and the designer can space the structural supports – the piles – without concern for how the space below will be used.

No one starts out wanting to build a pile-supported slab. A ground-supported slab is always preferable, and designers turn to piles only when the ground cannot adequately support the intended loads.

Floors on piles are always designed as suspended slabs with the piles as the structural supports. Some engineers also take into account the contribution of the ground between the piles, which usually offers some support even if it cannot carry the whole load. That can be tricky, however, because differential settlement may alter the support conditions over time.

Most pile-supported slabs are reinforced with two mats of steel – generally bars, but sometimes wire mesh. Some are post-tensioned. Recently a few designers have been experimenting with steel-fibre reinforcement for slabs on piles, either on its own or combined with bars.

It is usual to install the piles on a square grid, with piles spaced about 3 m (10 ft) apart. That keeps the spans short, simplifying the design and limiting deflection. But other patterns are possible, and often make sense where fixed, concentrated loads control the design. For example, some pallet-rack warehouse are designed with rows of piles directly under the back-to-back rack-legs, where the heaviest loads occur.

Choosing a floor type

Choosing a floor type is one of the most important decisions in designing a suspended floor. The choice depends on many factors, including:

- span (see Table 6.1);
- loading;
- need to control deflections;
- availability of formwork.

Designers must strike a balance between material and labour costs. Some elegant designs use little material, but require complicated forms that are costly to build. Simple one-way slabs, flat plates, flat slabs and slabs on metal deck are the workhorses of the suspended-floor world, largely because they are easy to build. When a more complicated floor type is needed it should be based on ready-made formwork if possible.

Local custom plays a big part in the choice of floor type. If a design is unfamiliar to the local builders, it often costs more to build.

The design process

Designers of ground-supported floors can choose between many design methods. In contrast, designers of suspended floors are practically obligated to follow their country's structural code. In America, the national code for reinforced concrete structures, including suspended floors, is ACI 318.

The UK equivalent used to be BS 8110, but Eurocode 2, *Design of Concrete Structures*, has replaced it. The British versions of Eurocode 2 appear as BS EN 1992-1-1:2004 (for general building design) and BS EN 1992-1-2:2004 (on design for fire resistance). Unlike the evolutionary changes that occurred from time to time in BS 8110 (and ACI 318), the switch to Eurocode 2 marks a sharp break in many ways. Nevertheless, British structural engineers have no choice but to learn it up.

A few general comments apply to all the design methods commonly used for suspended floors.

These are the main considerations:

- flexural strength of the floor (not the concrete);
- shear strength;
- control of deflection;
- control of drying-shrinkage and thermal-contraction cracks.

Most structural engineers start out by designing for flexural strength – the floor's ability to support its loads without structural failure. They then check the design for shear strength, deflection and crack control.

The last two considerations – control of deflection and cracks – typically fall under the heading of "serviceability" in books on structural engineering. I prefer to avoid that term in floor design, however, because it takes far more than control over deflection and cracks to make a floor serviceable. Real serviceability depends on the flatness, wear resistance and other factors that have little or nothing to do with structural design.

The process of structural design typically goes like this:

1. Collect information on the floor's live load.
2. Decide on the floor span, which may be controlled by the general building layout, or by the need to use the space under the floor.
3. Decide on the structural type (flat plate, waffle slab, etc.).
4. Design the floor for flexural strength, with a safety factor at least as great as required by the structural code.
5. Check the design for shear strength. Laminar shear is very unlikely to be a problem, but punching shear at columns can be.
6. Calculate deflections and ensure that they are within allowed limits.
7. Ensure that the floor has enough reinforcing steel to control drying-shrinkage and thermal-contraction cracks.

Sometimes there are shortcuts after step 3. Where the design relies on standardized products such as precast planks, the manufacturer normally supplies information on what size units to order for the desired combination of load and span.

Structural Design of Suspended Floors

Design considerations

In the rest of this chapter we look at some of the more important structural considerations:

- flexural strength;
- shear strength;
- deflections;
- drying-shrinkage and thermal-contraction cracks;
- loading;
- spans;
- concrete strength;
- reinforcing steel;
- prestress;
- formwork.

Flexural strength

The primary requirement for any suspended floor is flexural strength – the ability to resist bending stress. Every suspended floor needs flexural strength to support its own weight and all applied loads, along with a safety factor.

Bending induces both compressive and tensile stresses in various parts of the floor. Because of this, flexural strength involves both compressive strength and tensile strength.

Plain concrete has some inherent flexural strength (see Chapter 7), but the designers of suspended floors rarely give it much thought. (It does matter in many ground-supported floors which are designed on different principles.) The important thing is the flexural strength of the whole floor, which consists of concrete and steel working together. We rely on the concrete, to resist the compressive stresses, and on the steel to resist the tensile. Thus for concrete, the critical property is compressive strength, while for steel it is tensile strength.

The structural codes allow several methods for determining flexural strength. But with any method, these are the main variables to consider:

- the type of floor (flat plate, one-way, etc.);
- the span, or distance between structural supports;
- continuity across structural supports;

- compressive strength of the concrete;
- tensile strength of the reinforcing steel;
- slab thickness;
- location and amount of reinforcing steel.

A good design combines these variables in a way that is safe, economic and in conformity with the applicable structural code.

Shear strength

The structural codes require designers to check suspended floors for shear. There are two forms to consider, laminar (or horizontal) and punching.

Laminar shear occurs when an upper layer of the floor slab tries to slide relative to a lower layer. Imagine a floor made of sheets of paper piled high. As you press down on the floor, each sheet would slide slightly on the one beneath it. That sliding is laminar shear.

Floor slabs almost never fail in laminar shear, because they are so thin. Shear stress develops, but the concrete is strong enough to resist it. In contrast, the thick beams that support some floors can develop shear stress that exceeds concrete's modest shear strength. The usual solution is vertical reinforcing bars called stirrups.

Punching shear is different matter, and sometimes controls the floor design. The worst case occurs near the columns in a beamless floor (flat plate or flat slab). Designers deal with punching shear by adding reinforcement or by thickening the slab over the columns (turning a flat plate into a flat slab, for example). Thickened sections over columns are sometimes called shear panels.

Deflection

This is the bending of a suspended floor under load. It is usually expressed as the ratio of greatest vertical displacement to span (Figure 6.x). All suspended floors deflect. Large deflections create serious problems for floor users, and even modest, normal deflections sometimes cause trouble.

The main problem is that deflection makes the floor surface less level. In an industrial building, a badly defected floor may not be level enough for some materials-handling vehicles. In an office, it may be hard to set up movable partitions on a deflected floor.

Another problem is psychological: users lose confidence in a floor with large obvious deflections. They may have good reason to be concerned, for severe deflection can be a sign of imminent collapse. But even when a structural engineer certifies that a sagging floor will not fall down, people may not trust it.

How much deflection is too much? The traditional rule states that deflection from live load should not exceed 1/360 of the span. By this rule, a floor spanning 10 m (33 ft) could deflect by 28 mm (1.1 in). That much deflection would not be acceptable in some modern warehouses and offices. Some floors have been designed for deflections as small as 1/2000 of the span. On the other hand, deflections greater than 1/360 are tolerable for some applications.

Structural codes provide guides for predicting deflection. For some standard designs, the structural engineer is allowed to specify a minimum thickness instead of calculating the deflection.

Following the code does not guarantee success, however. One problem is that some users require a more level floor than the codes aim for. A more fundamental problem is that deflections vary hugely and cannot be predicted with much accuracy.

Cracking is the main reason we cannot accurately predict deflection. Uncracked concrete deflects predictably, but we can rarely count on getting it. Reinforced-concrete design is based on the premise that concrete has negligible tensile strength and will crack in zones of tension. But in fact concrete has some tensile strength. Sometimes it cracks in tension, and sometimes it does not. An uncracked slab will deflect less than a slab with many cracks.

Suppose two floors are built to the same design for the same usage, but under different conditions. The first job takes place in cool weather. It is ahead of schedule, so forms and shores stay in place for several weeks. The concrete ages for many months before any substantial loads are put on it. The second job takes place amidst a heat wave. It is behind schedule, so forms and shores are snatched away at the earliest moment allowed under the contract. Following trades pile construction materials on the floor and the user moves in as soon as possible.

The first floor would almost certainly have fewer cracks and less deflection than the second. Both floors would have the same ultimate strength. If loaded to the point of collapse, both would have cracked and their tensile strength would be controlled by the reinforcing steel. But in normal use the first floor, with fewer cracks, would be stiffer.

If adherence to the structural codes does not always protect against large deflections, what can we do? For some floor uses, nothing need be done because large deflections cause no trouble. No one is likely to notice deflection in a multi-storey car park. For floors where large deflections would be objectionable, there are several ways to improve the odds.

The choice of structural type affects deflection. Two-way beamless floors and slabs on metal deck are generally not good choices where deflection must be tightly controlled. Slabs on beams work better. One-way joist systems and two-way waffle slabs have the best record for small deflections.

Another way to control deflection is to use prestress instead of steel reinforcement. Prestressed floors do not necessarily deflect less; they can be designed for large or small deflections. But they deflect more predictably, because they crack less.

We can also reduce deflections by overdesigning a floor – making it thicker or using more reinforcement than required by the structural code.

One almost painless way to reduce deflections is to protect the floor from early stresses that cause cracks. It helps to keep the forms and shores in place for as long as possible, and to prohibit or limit early construction loads.

Drying-shrinkage and thermal-contraction cracks

Chapter 13 describes in detail drying-shrinkage and thermal-contraction cracks.

The structural codes contain minimum steel requirements to control cracks. The requirements apply both parallel and perpendicular to the span. A one-way slab might be reinforced in only one direction for flexural strength, but must meet the minimum requirements for crack-control reinforcement in both directions.

The code requirements do not suffice for every floor. Where good control of cracks is important, some designers use more reinforcement than the code requires.

If properly designed and built, prestressed floors have very few cracks.

Suspended floors rarely include the stress-relief joints commonly used to control cracks in ground-supported slab. Some designers call for sawing crack-control joints in slabs on metal deck, but even there the practice is controversial.

Loading

Chapter 5 defines the types of floor loading.

Suspended floors are normally designed for nominal live loads expressed in kN/m² (lb/ft²). This is a practical, if not completely accurate, way to describe the light loads most suspended floors are called on to support.

When a suspended floor must support heavy industrial loads, it should be designed for the specific loading pattern and not for a nominal loading.

Span

The span, or distance between structural supports, has a big effort on the floor's structural design. Sometimes it dictated the type of slab used. Table 6.1 shows the maximum practical spans for various structural types.

Spans are usually determined by factors independent of the floor. For most elevated floors, the spacing of structural supports is controlled by the need to use the space under the floor. Closely spaced columns in a warehouse or factory might simplify floor design, but would make the room below harder to use.

The situation is completely different, however, for ground-level suspended floors built on piles. Because designers have complete control over the pile spacing, they often design these floors with short spans of only 3 m (10 ft) or so. When a piled floor is designed to support concentrated loads, it is sometimes practical to put the piles directly under the loads. Though uncommon, this practice can greatly lower the cost of the floor.

Table 6.1 Maximum practical spans for various floor types

Type of Floor	Maximum Span
Slab on unshored metal deck	3 m (10 ft)
One-way slab	6 m (20 ft)
Slab on shored metal deck	6 m (20 ft)
Flat plate	9 m (30 ft)
Flat slab	10 m (34 ft)
One-way joist slab	14 m (45 ft)
Waffle slab	15 m (50 ft)

The maximum spans listed here are not absolute limits. Longer spans are possible with every type, but may not be economical.

Concrete strength

Chapter 7 covers this subject in more detail. For a suspended floor, the critical property is the concrete's compressive strength. This is usually between 10 MPa and 40 MPa (2500 psi and 5000 psi). Although stronger concrete can be made, it is of little use in suspended floors. Working in Canada, Salinas (1980) showed that the cheapest floor designs had the lowest strength concrete and the highest strength reinforcing steel.

Concrete's shear strength also matters, but is never specified or measured directly. The structural codes contain empirical rules for dealing with shear. Where the calculated shear stress exceeds the concrete's estimated shear strength, the solution is never to make the concrete stronger (though that might work), but to add reinforcing steel perpendicular to the direction of shear.

Concrete's tensile strength is rarely considered in the design of suspended floors. Reinforced-concrete design is based on the principle that concrete has zero tensile strength, although that is not strictly true.

For some suspended floors, the concrete strength is controlled by factors unrelated to structural design. Designers may specify high strength to achieve other goals such as good surface finish, wear resistance and resistance to chemical attack.

Reinforcement

Most suspended floors contain steel reinforcement, which serves three purposes:
- to resist tensile stress induced by bending under load;
- to resist shear stress (mainly from punching shear at columns);
- to limit the width of drying-shrinkage and thermal-contraction cracks.

One of the designer's biggest jobs is to design and detail the reinforcing steel. The main decisions involve the amount and location of the steel. But there are some other issues that need to be considered, including:
- type of reinforcement (bars or mesh);
- tensile strength of the steel;
- concrete cover.

Designers can choose between two types of reinforcement: individual bars and wire mesh.

Individual bars are steel rods, available in various diameters and lengths. Nowadays the bars are invariably deformed, which means that bumps have been formed on them to help anchor them within the concrete. The advantage of bars is that they are easily tailored for specific needs. Though bars are only available in certain sizes, the spacing between bars is infinitely variable and can be adjusted to provide the needed amount of reinforcement. The disadvantage is that bars require time and labour to install.

Wire mesh, also called steel fabric, is a grid of smooth steel bars, welded at all intersections. Wire mesh is made in a factory and delivered to the building site in flat sheets or rolls. It is available with various bar sizes and spacings, but the options are limited compared to those available with individual bars. The advantage of mesh – a big advantage – is that it saves time and labour on site. The disadvantage is that it uses material less efficiently. Because mesh is only available in certain sizes, floors with mesh reinforcement sometimes wind up having more steel than necessary in at least one direction.

The steel's tensile strength matters because it determines how much steel is needed to resist the floor's tensile stresses. Stronger steel brings advantages, but there is no point in specifying higher strength than the steel manufacturers offer. In the UK, common reinforcing steel has a tensile strength (yield strength) of 460 MPa (67 000 psi). The American standard is called Grade 60, with tensile strength of 410 MPa (60 000 psi).

Cover is the distance between reinforcing steel and the concrete surface. If we consider only structural strength, the most efficient design has the least of cover on the tension side of the floor. For that reason structural engineers normally use the minimum cover allowed by the structural codes, which is as little as 20 mm (3/4 in) in some circumstances.

Many floors need more than the minimum cover, however, since structural strength is seldom the only consideration. Steel – especially large-diameter bars – can interfere with finishing if it lies within 40 mm (1.5 in) of the surface. ACI 302 suggests at least 50 mm (2 in) of cover is desirable to prevent plastic-settlement cracks. Floors that support wire-guided vehicles need extra cover to prevent interference. Wire-guidance suppliers typically ask for about 50 mm (2 in) of cover, but some demand as much as 100 mm (4 in). All these requirements for extra cover apply to the top of the floor. There is seldom any problem at the bottom.

When inadequate cover causes construction problems, builders are tempted to set the reinforcing steel lower than the drawings shown. But that is risky. The position of the reinforcement is an important part of the structural design, and should only be changed by the responsible structural engineer.

Prestress

Most suspended floors are structurally reinforced, but some are prestressed instead, using stretched steel cables. Prestressing creates compressive stresses in concrete to offset the tensile stresses that result from shrinkage, thermal contraction and loading. If done right, prestress can effectively eliminate all tensile stresses in a concrete floor.

Prestressing steel differs from reinforcing steel. The typical prestressing tendon consists of six wires twisted around a straight core. The steel is very strong, with a yield strength of about 1860 MPa (270 000 psi).

Two kinds of prestressing are used in suspended floors: pre-tensioning and post-tensioning. The prefixes can get confusing, I admit. The pre in prestressing refers to the fact that stress is induced before the slab is loaded. The pre in pre-tensioning refers to the fact that the tendons are tensioned (stretched) before the concrete is cast around them. The post in post-tensioning refers to the fact that the tendons are tensioned after the concrete has hardened.

Pre-tensioning is seldom practical on the job site, but is widely used for factory-made items such as hollowcore planks. Bare tendons are stretched over a strong frame, which is then filled with concrete. After the concrete has set, the tendons are released, putting the concrete in compression.

Post-tensioning is more practical on site. The tendons are sheathed to isolate them from the wet concrete, and are left untensioned during concrete placement. After the concrete has set and grown strong enough to take the stress, the tendons are stretched and connected through wedges to steel anchors near the slab edges. Most designs call for the tendons to be draped or profiled. That means the tendons curve up and down to counteract bending moments. (In contrast, ground-supported floors usually get straight tendons.) The tendons are normally highest over the structural supports, and lowest at mid-span.

Post-tensioning tendons come in two varieties: unbonded and bonded. An unbonded tendon is greased and covered with a plastic sheath. It applies force to the concrete through the anchors at each end. A bonded tendon is installed within a rigid tube. After stressing, grout is pumped into the tube, tying the stretched tendon to the concrete all along its length. Unbonded tendons are well protected from corrosion, an important consideration in multi-storey car parks. Bonded tendons produce a finished product very similar to that from pre-tensioning, without the long-term stress concentration at the end anchors.

Prestress of all kinds offers at least four advantages over reinforcement.

First, it uses less steel. Even though prestressing steel costs more than an equal weight of reinforcing steel, this higher price is usually more than offset by the fact that much less steel is needed. Prestressed concrete was first developed as the solution to a steel shortage.

Second, prestress allows thinner slabs. With prestress, the whole thickness of the slab is effective and can be used in structural calculations. But with reinforcement, the slab's thickness is the distance from the concrete surface on the compression side to the steel reinforcement on the tension side.

Third, prestress prevents most cracks. Reinforcement does not prevent cracks; it can only limit the width of cracks after they have occurred. Prestress, in contrast, really does prevent cracks by preventing the concrete going into tension.

Fourth, prestress give us better control over deflection. Prestressed floors do not necessarily deflect less, but their deflections are easier to predict accurately.

Formwork

Formwork is the temporary structure that contains and supports the fresh concrete. It is the mould into which a suspended floor is poured. It consists of the forms themselves, along with props, hangers and other hardware needed to hold the forms in place.

Soffit forms define the bottom of the floor slab and must support the fresh concrete's weight. They determine how the underside of the floor will look – an important consideration in multi-storey car parks and other buildings where the soffit is left exposed to view. Slabs on metal deck use the deck as the soffit form.

Side forms define the sides of the floor slab and are used to make construction joints. Side forms for floors need no great strength, because most floors are not thick enough to develop much hydraulic pressure. But they do need to be set accurately, because they have a big effect on the floor's flatness and levelness.

In the old days, formwork was built on site out of timber and plywood. That's still done, but it costs a lot in labour and materials. Increasingly builders rely on reusable, ready-made steel forms.

Formwork can account for more than half the cost of a suspended floor. A simple design that allows cheaper formwork will often reduce total cost, even if it uses more concrete than a more complicated design. Flat plates and simple one-way slabs are the easiest types to form up.

Questions are often raised as to when forms can be removed. The striking of side forms has no structural effect and can be done soon after pouring concrete – usually within 12–18 hours, according to Blackledge (1977, p. 13). In contrast, the striking of soffit forms and the props (if any that support them) does have structural effect and must be done with care. Specifications typically require that the soffit forms be left in place till the concrete has reached a certain strength. Where the schedule demands the earliest possible form-striking, it is essential to determine the concrete's strength from field-cured cubes or cylinders. The usual laboratory-cured samples can mislead. Where good control over deflection and cracking is needed, it helps to leave the forms up longer than required for structural safety.

In America, ACI 117 specifies that floor levelness tests (the F_L readings) must be finished before soffit forms and props are removed.

Part III

Concrete for Floors

7

Properties of Plastic Concrete

This chapter deals with those properties of plastic (not yet hardened) concrete that affect floor construction. Concrete is plastic for just a few hours, but the events of those hours determine how a floor performs over its whole lifespan. For properties of hardened concrete, see Chapter 8.

These are the properties of plastic concrete that matter most in floors:

- workability;
- finishability;
- bleeding;
- setting time;
- plastic settlement.

Workability

This is the ease with which plastic concrete can be compacted and moved. Concrete of high workability is called wet or loose. Concrete of low workability is called dry or stiff. Because the commonest measure of workability is the slump test, people often define workability in terms of slump; high slump means high workability.

Newer British standards call workability consistence. I suppose we will have to get used to that, but I intend to use the older term as long as it remains understood. Calling it consistence risks confusion, since the term can also mean constancy – also a desirable property for concrete to have, but quite a different one.

The American Concrete Institute defines workability as "that property of freshly mixed concrete...which determines the ease and homogeneity with which it can be mixed, placed, compacted, and finished". Implying that finishability is simply one facet of workability, the ACI definition serves well enough for much concrete work, but not for floorlaying. In a floor, workability and finishability are distinct properties. Workable concrete is easy to place, compact and strike off. Finishable concrete is easy to straightedge, float and trowel. Workability is important in all kinds of concrete construction, but finishability matters only in floors and other objects with large surfaces not cast against forms. While both workability and finishability are properties of plastic concrete, they apply at different times. Workability is essential when concrete is mixed and placed. Finishability comes into play later when the floor is floated and trowelled.

A controversy

Workability generates more controversy than almost any other subject in floor construction. Its only serious competitor in controversy is the matter of crack control in ground-supported slabs.

Workability is a good thing, but you can have too much of it. To make concrete more workable, we add water or an admixture. If we buy workability by adding water, the price includes lower strength and increased drying shrinkage. If we buy workability by adding an admixture, the price consists of the monetary cost (which can be substantial, especially for superplastizers) and an increased risk of inconsistent concrete and problems in finishing.

Because high workability is both desirable and costly, every concrete mix is a compromise. That alone would hardly be controversial; every floor design is full of compromises. What makes this subject different is sharp disagreement over what the compromise should be.

Many specifications limit slump to 50mm (2in) or 75mm (3in). In contrast, the main American guide to floor construction, ACI 302, allows up to 125mm (5in) for all floors except concrete toppings (Figure 7.1). In the UK, the Concrete Society (2003, p. 71) goes even further, recommending a target slump of up to 150mm (6in). Since that is a target, not a maximum, individual batches could presumably have an even higher slump without going out of tolerance.

Figure 7.1 A slump of about 100–125 mm (4–5 in), as shown here, often works well in slab construction

Often it comes down to a conflict between designers and floorlayers. The designers specify very stiff mixes; the floorlayers prefer wetter mixes and use them whenever they can. Designers believe that floorlayers care nothing about concrete quality, while floorlayers are equally convinced that designers know nothing about finishing concrete. The conflict often descends to outright deception. Some designers specify absurdly low slumps, expecting the floorlayers to cheat. Some floorlayers oblige them.

I confess a bias in favour of the floorlayers and higher workability. It is true that some floorlayers use mixes that are much wetter than they need to be. Nevertheless, a good floorlayer knows better than anyone else what degree of workability is right for the tools at hand. Forcing workers to use concrete that is too stiff does not make a better floor.

Measuring workability

No single test can handle the whole range of concrete workability. Table 7.1 shows six tests used on concrete for floors.

Table 7.1 Workability tests

Test	Workability	Effective Range as Measured by Slump Test	
		mm	in
Inverted slump cone*	Very low	0–50	0–2
Compacting factor	Very low	0–50	0–2
Vebe	Low	0–60	0–2½
Slump	Medium to high	25–150	1–6
K-slump	Medium to high	50–175	2–7
Flow table	Very high	Over 150	Over 6

*Only for concrete with fibres

Only the slump and K-slump tests provide useful results within the range of workability found in most floor pours. The compacting-factor and Vebe tests are suitable for the very stiff mixes used with road paving machines. Such machines are occasionally used to lay floors. At the other extreme, the flow-table test is usable on flowing concrete made with superplasticizer. The inverted-slump-cone test is meant only for stiff mixes that contain fibres.

In the UK, BS EN 12350 covers the slump, Vebe and flow-table tests. The compacting-factor test appears in BS 1881, but that is on its way out and the test may become an orphan.

In America, ASTM publishes a separate standard for each test: ASTM C 143 for slump, ASTM C 1362 for K-slump, ASTM C 1170 for Vebe consistometer and ASTM C 995 for the inverted slump cone.

Despite valid arguments in favour of each method, it is actually quite rare to see anything other than the good old slump test (Figure 7.2).

The slump test

This is by far the most common test for workability. It is so common that many people treat slump as a synonym for workability.

To test slump, you fill a mould with fresh concrete. The mould is a truncated cone 300 mm (12 in) high, open at both ends, with its wide end resting on a smooth, level base. After filling the mould and compacting the concrete with a steel rod, you lift the mould slowly. The de-moulded concrete then slumps under its own weight. The test result is the difference between the original height of the cone and the height after de-moulding. When we say a slump is high, we mean that the cone settled by a large amount.

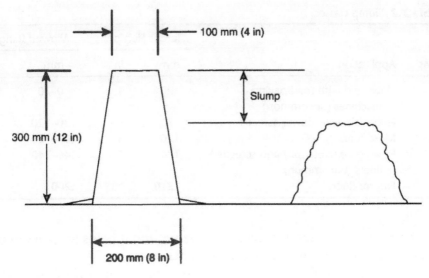

Figure 7.2 Slump test

BS EN 12350-2 and ASTM C 143 describe the test, which differs slightly between the UK and North America. The most important difference is that UK testers measure to the highest point on the slumped concrete, while Americans measure to the centre of the displaced top. The difference tends to make American slump results a few millimetres higher.

Slump specifications also differ between the UK and America. The British traditionally specified slump in multiples of 25 mm, which corresponded closely to the whole-inch values used in pre-metric days. European standards have changed all that. We are now supposed to specify and report slumps in multiples of 10 mm. A system of slump classes (see Table 7.2) has come into use. Each class has its own target slump in the middle of a permitted range.

American standards offer more choices. No slump classes exist, and ACI 117 allows slump to be specified in three ways: as an absolute maximum, as a range (also absolute) or as a target. If specified as a target, slump can vary up and down by a limited amount. Where the target is 100 mm (4 in) or less, the permitted variation is ±25 mm (±1 in). Where the target is above 100 mm (4 in), the permitted variation grows to ±40 mm (±1½ in).

I recommend specifying a maximum slump rather than a range or a target, even though that departs from the latest British standards. On the job site it is always the upper limit that people pay attention to. No one worries much about rejecting concrete that arrives too stiff, and you can always increase workability by adding a little water. But everyone understands it

Table 7.2 Slump classes

Class	Application	Target Slump		Allowed Range	
		mm	in	mm	in
S1	Floors laid with road-paving machines (uncommon)	20	1	0–70	0–3
S2	Floors laid with power screeds	70	3	30–120	1–5
S3	Most floors	120	5	80–180	3–7
S4	Flowing concrete for hand-screeded floors (uncommon)	180	7	140–240	5–9
S5	Not for floors	≥210	≥8	≥200	≥8

is important not to exceed a maximum value, and it helps to spell out that maximum value in the contract documents.

Typical slumps in floor construction range from 50 mm (2 in) to 150 mm (6 in). Slumps near the lower end are used where concrete is struck off with power equipment. Slumps near the higher end are used where concrete is struck off by hand, and where concrete is pumped.

Instead of slumping, the cone of concrete sometimes breaks into two parts with the top part sliding down an inclined plane. This is called a shear slump. It may point to sloppy testing or a bad, non-cohesive mix, but some perfectly serviceable mixes tend to produce shear slumps.

The slump test is both useful and very widely used. On some jobs it is the only test to which concrete is put. But we need to keep in mind its limitations.

The slump test does not measure strength. While some correlation exists between slump and strength, it is unreliable. If two batches are identical in composition except for water content, the wetter mix will have the higher slump and, under most conditions, the lower strength. But that statement assumes a condition rarely met. Many factors other than water content affect slump. High-slump concrete can be strong, and low-slump concrete can be weak. Only direct tests strength tests on the hardened concrete reliably tell strong concrete from weak.

Another limitation is that the slump test works only over a small range of workability. We cannot rely on the standard test for slumps lower than 25 mm (1 in) or higher than about 150 mm (6 in). That may seem a wide range – 150 mm is six times 25 mm – but many concrete mixes fall outside it. The stiff mixes used with paving machines often produce slumps under 25 mm (1 in), while flowing concrete made with superplasticizer often tests above 150 mm (6 in).

Yet another limitation is the slump test's sensitivity to operator technique. Repeatability is none too good even when operators follow every rule to the letter. If the operator fails to follow the standard procedure, the results can be off by as much as 75 mm (3 in). And unlike some other concrete tests, slump tests are often made by untrained workers using non-standard tools. I once saw a slump cone that had been run over by a van and hammered back into shape.

As long as those limitations are understood, the slump test can provide useful information. The main value of slump testing is to detect irregularities in the concrete mix. If the concrete batcher makes a serious mistake, it usually shows up as a large change in the slump. By slumping every batch, or at least any batch that looks abnormal, you can often identify a batching problem before the concrete has been made part of the floor.

Arguments arise over the location and timing of slump tests. Some specifications require that samples be taken from several parts of a batch. ASTM C 172 requires the taking of samples "at two or more regularly spaced intervals during discharge of the middle portion of the batch", while forbidding samples "from the very first or last portions of the batch discharge". The American standard also states that the concrete shall not be tested till all the water has been added to the mix. British requirements are more liberal, but still forbid sampling the first part of the batch.

These requirements make sense if the goal is simply to get a fair sample of the concrete. But if the goal is to identify mis-batched concrete in time to reject it, or to adjust workability by adding water, we need to test slump earlier. That means sampling the first part of the batch discharged from the truck or mixer. There is a chance that the sample will be unrepresentative, but there is no other way to check the mix in time to do any good.

Similar logic applies to the sampling of pumped concrete. Because concrete loses workability during pumping, builders often wish to sample the mix at the discharge end of the pipeline. Their wish is understandable, but to grant it means that a substantial volume of concrete is already in the pumping system by the time of the first slump test. It is better to sample the concrete before it enters the pump. If slump loss during pumping is a problem, the specification should allow a higher slump at the start.

The K-slump test

The slump cone is easy to use, but that fact has not stopped people from developing simpler and faster devices for measuring workability.

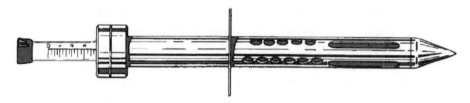

Figure 7.3 K-slump tester

The K-slump tester (Figure 7.3) is one such device. The K-slump test is also known as the flow test – not to be confused with the flow-table test.

It consists of a steel tube with a measuring rod that slides within it. The tube is perforated to let cement paste enter, and pointed at one end so it can be pushed into the concrete. A flange ensures that the tester is inserted to the right depth. The measuring rod is of light plastic, calibrated in millimetres from 0 to 100.

To test K-slump, you stick the tester into fresh concrete. A hook holds the measuring rod above the concrete surface. You then wait 60 s, during which time mortar flows into the tester through the perforations. Higher workability means more leakage. When the minute is up, you lower the measuring rod gently till it rests on the mortar that entered the tester. The measuring rod then slows the test result, in millimetres, on a scale that runs from 0 to 100 mm. As with the slump test, higher readings mean greater workability.

In the early days of the K-slump test, results were read in centimetres and the inventor claimed that K-slumps in centimetres were roughly equal to slumps in inches. In other words, a mix that gave a K-slump of, say, 4 cm could be expected to give a regular slump of 4 in (100 mm). However, the correlation between K-slump and slump has proven unreliable, and the current standard for K-slump, ASTM C 1362, does not even mention it.

The K-slump tester offers three advantages over the slump cone. It is slightly easier to use; it is faster and it can be used almost anywhere – in a bucket, in a wheelbarrow or in the chute of a concrete truck. These advantages combine to make the K-slump test particularly useful on a fast-paced job where you want to check workability on every batch but cannot delay the pour waiting for slump tests.

The chief drawback to the K-slump tester is that people are unfamiliar with it, despite the existence of an ASTM test method describing and standardizing its use. If you propose to use the K-slump test for acceptance testing, you had better expect some resistance, not least from the ready-mix supplier.

The compacting-factor test

Neither the slump cone nor the K-slump tester gives usable results on mixes of low workability. Such mixes are not common in floor construction, but are seen on a few jobs where big industrial floors are laid with road-paving machines. The compacting-factor test is one way to measure workability on those jobs. The Vebe test, described below, is another.

It compares the weights of two equal volumes of concrete, one of which has been thoroughly compacted by ramming or vibration. The other has been only partially compacting by falling through two hoppers. The test result, called the compacting factor, is the ratio of the weight of the partially compacted concrete to the weight of the thoroughly compacted concrete.

Typical results range from 0.75 to 0.95, with higher numbers denoting greater workability. A compacting factor of 0.92 corresponds to a slump of about 50 mm (2 in).

The Vebe consistometer

Like the compacting-factor test, this measures the workability of stiff mixes such as those used with road-paving machines. It can be used on site, but is better suited to the laboratory.

The test device is a cylindrical mould mounted on a vibrating table. To make the test, you fill the mould with fresh concrete and vibrate it till it looks fully compacted, as shown by mortar forming around the sample. (ASTM C 1170 and BS EN 12350 provide details.) The test result is the elapsed time to full compaction, measured by stopwatch. Results range from 3 to 30 s, with longer times denoting lower workability.

ASTM C 1170 claims the test is usable with aggregates up to 50 mm (2 in), but not everyone agrees. Murdock, Brook and Dewar (1991, p. 96) say "the results are of doubtful value" where aggregates exceed 20 mm (3/4 in).

Table 7.3 shows the rough correlation between Vebe time and slump.

The flow-table test

None of the tests described so far is very useful for flowing concrete made with superplasticizer. If the concrete has a design slump of more than 150 mm (6 in), it is best tested on a flow table.

Table 7.3 Correlation of slump and Vebe time (Cement Association of Canada)

Description of Workability	Slump mm	Slump in	Vebe Time Seconds
Very low	0–10	0	Over 12
Low	10–30	Under 1	6–12
Normal	30–60	1–2	3–6
High	60–180	2–7	0–3

The test involves these steps:

1. Place a shortened slump cone in the centre of a 700 mm by 700 mm (28 in by 28 in) table.
2. Fill the cone with concrete in two layers, tamping each layer 10 times with a wooden rod.
3. Wait 30 s.
4. Lift the cone mould and set it aside.
5. Lift the table 40 mm and drop it. Do this 15 times in 15 s.
6. Measure the mean diameter of the concrete puddle.

The test result is the diameter of the puddle, with greater diameters denoting higher workability.

Inverted-slump-cone test

This test was devised for concrete that contains fibres. The addition of fibres, steel or plastic, can substantially reduce slump as measured in the usual way. However, the real decline in workability may be less than the low slump results would suggest. The inverted-slump-cone test offers a way around that problem.

The test relies on a standard slump cone, but similarities to the standard slump test end there. You support the cone upside down (wide end up) over a bucket. You fill the cone with fibre-reinforced concrete. If the concrete is stiff enough, it stays in the cone and the test goes on. If it flows right out of the cone's open bottom, then it is too workable for this test and you can switch to the standard slump test. Provided the concrete stays in the cone, you then insert a poker vibrator of 25-mm (1-in) diameter. The vibration loosens the concrete and makes it fall out of the cone. The test result is the elapsed time, measured by stopwatch, from the insertion of the vibrator to the emptying of the cone.

According to ASTM C 995, if the elapsed time is under 8 s, the test is invalid and the regular slump test should apply instead.

How much workability?

We have discussed ways to measure workability, but we are not much closer to deciding how much of it we really need. No single answer will suffice, because the answer depends on many factors, including but not necessarily limited to:

- temperature (High temperatures call for high workability.);
- protection from the elements (Walls and roof allow lower workability.);
- concrete placing method (Pumping requires high workability.);
- strike-off method (Manual methods need higher workability than do vibrating screeds and laser screeds.);
- finish (Dry shakes are hard to apply on concrete of low workability.).

Since we have to start somewhere, I suggest that a maximum slump of 120 mm (5 in) is about right for the average floor laid under average conditions.

In practice, the workability is often controlled by the need for finishability. With enough effort, we can always place concrete of low workability. But we may not be able to finish it.

Finishability

This is the ease with which concrete can be straightedged, floated and trowelled. In floor construction, finishability matters even more than workability. You can, with effort, place concrete of low workability. But you cannot put a good wearing surface on concrete of poor finishability. For smooth trowelled floors, finishability is one of the most essential concrete properties.

Despite its importance, finishability has not received much study. Most of our knowledge comes not from formal research, but from the experience of workers on site.

There is no standard test for finishability. In its absence, we can only rely on knowledge of those factors that make concrete finishable, and on workability tests (mainly the slump test) that indirectly and imperfectly indicate finishability. Because the only true test is actual site experience, we should stay ready to change mixes that are hard to finish.

Finishability is a property of the surface mortar that forms on the top of a concrete slab. Because of this, finishability is affected by the components of that mortar. Water, cement and fine aggregate all appear in the surface mortar, and all affect finishability. Coarse aggregate does not matter much because it is not part of the mortar. (Coarse aggregate has an indirect effect, however, in that its size affects the amount of cement needed for finishability.)

Water and finishability

Water makes concrete finishable by lubricating the cement and aggregate particles in the surface mortar. The amount of water needed for finishability is much greater than the minimum needed to hydrate the cement.

The amount of water in the surface mortar depends mainly on the water content of the original concrete mix, but that is not the only factor. Bleeding wets the surface mortar. Other processes – absorption by the aggregates, chemical reaction with the cement and evaporation – reduce the amount of water available to lubricate the mortar.

No one can predict exactly how much water a concrete mix needs to ensure finishability, but we have to try to get it right. Use too much water, and excessive bleeding makes finishing harder. Use too little, and the surface dries out and becomes impossible to finish. Most of the time, adding enough water to produce a slump of 75–125 mm (3–5 in) will make a finishable mix, provided no other factors interfere. The lower end of that slump range works in cool, damp weather. The higher end is more suitable for hot or dry conditions.

If the surface mortar dries out too soon, floorlayers sprinkle it with water. Many specifications forbid the practice, but it goes on anyway. It is better to sprinkle a small amount of water on a floor than to leave the surface unsealed and porous.

Though an occasional sprinkle of water is nothing to be afraid of, the use of more than an occasional sprinkle points to a problem that needs to be corrected. If the problem is poor finishing habits, then the solution is to change those habits. In most cases, however, the solution is to use more water at batching (that is, to raise the slump), or to reduce the rate of evaporation.

Superplasticizer cannot replace water where finishabilty is concerned. Superplasticizer affects workability more than finishability. By the time finishing takes place, superplasticizer has lost its effect.

Cement and finishability

More cement means greater finishability, provided water is also added to maintain the water–cement ratio. To ensure a high cement content, some designers specify a minimum concrete compressive strength – typically 40 MPa (4000 psi) – even if the structural design would allow less. That usually works, but it creates the risk that a concrete supplier will develop a mix that attains high strength with relatively little cement. The safe approach is to specify strength for strength's sake, and also to specify a minimum cement content for finishability.

Deacon (1986) recommends a cement content of at least 300 kg/m^3 (560 lb/yd^3) for concrete that forms the floor's wearing surface. ACI 302 makes a similar recommendation, but tailors the minimum cement to the maximum size of the coarse aggregate. The bigger the aggregate, the less cement is needed for the same finishability.

Fine aggregate and finishability

For a good finish, concrete must contain enough fine aggregate to cover the floor surface with mortar. An undersanded mix – one without enough fines – is always hard to finish. There are no standard guidelines to the amount of fines needed, but most ordinary mixes contain enough.

The shape of the fine aggregate is also important. Rounded particles are better than sharp, angular ones – though sharp aggregate may produce higher concrete strength. Designers rarely specify or even consider this property, but it is worth investigating when other measures fail to produce a finishable mix.

Ways to improve finishability

When a mix is hard to finish, one or more of the following changes may improve the situation:
- Add water, but beware the effect on strength and drying shrinkage.
- Use more cement.
- Use more fine aggregate.
- Switch to a round-grained fine aggregate.
- Protect the fresh concrete from drying winds and sunshine.

- Use fog sprayers to keep moist air around the concrete. The goal is not to deposit liquid water on the floor, but to stop the floor surface's drying out by raising the air's humidity. The sprayers release a very fine mist.
- Spray the concrete with a anti-evaporation film to reduce evaporation.

Bleeding

This is the process by which water rises to surface of a newly placed floor slab. Concrete bleeds because water is the lightest component of the mix (except some aggregates used in lightweight concrete). The heavier components tend to settle, with water rising to the top (Figure 7.4).

When bleed water rises faster than it can evaporate into the air, it accumulates on the floor surface. In extreme cases a sheet of water covers the whole floor. In milder cases the water ponds in isolated spots. The extent to which bleed water appears on the floor surface depends both on the

Figure 7.4 Bleeding

bleeding rate – the rate at which water rises through the concrete – and on the evaporation rate. In cold, damp weather almost any concrete mix will produce visible bleed water. Under hot, dry, windy conditions, the floor surface may stay dry even with a wet, high-bleeding concrete mix.

A little bleeding does no harm, and may do some good by keeping the surface moist enough for easy finishing. If concrete did not bleed at all, it would be harder to finish in dry air. Dry shakes, which work by soaking up moisture from the concrete, would be all but impossible to apply. You could argue that the ideal concrete mix is one that bleeds just enough to offset the loss of water by evaporation.

Although a little bleeding helps, a large amount does not. The main risk from excessive bleeding is that concrete finishers will float or trowel the floor while bleed water covers its surface. The floating or trowelling mixes water back into the concrete, creating a weak layer with an abnormally high water–cement ratio. Results may include severe crazing, delamination and poor resistance to wear.

The standard and usually effective way to prevent those defects is to follow one simple rule: never float or trowel concrete while bleed water is present. Wait for the water to evaporate.

While that is a good rule, very heavy bleeding makes it hard to follow. In some cases the water evaporates so slowly that by the time it has disappeared, the concrete has set and is too hard to finish. In less severe cases the floor may turn out well, but the delayed finish raises labour costs.

Here are three basic ways to deal with excessive bleeding and let finishing proceed without long delay:

- remove bleed water mechanically;
- increase the evaporation rate;
- change the mix design to reduce the bleeding rate.

The traditional mechanical method is to drag the water off with a rubber hose. Nowadays finishers may use a highway straightedge instead of a hose, though some people believe the straightedge, weighing more than a hose, forces too much water back into the concrete surface. Both tools work best on narrow slabs.

Another mechanical method is vacuum dewatering. A mat, consisting of a mesh layer under an impervious plastic sheet, is rolled out over the concrete surface. Vacuum is then applied under the plastic sheet, sucking out water. Finishing can usually start as soon as the mat is removed.

Yet another mechanical method relies on dry cement to absorb excess water. The cement is not applied directly to the floor, because it would then bond to the wet concrete. First dry hessian (Americans call it burlap) is laid over the floor. Then dry cement is swept over the hessian. After the cement has become thoroughly wet, the hessian is rolled up and finishing can begin.

Increasing the evaporation rate can solve the problem of excessive bleeding, but is not practical on every project. On outdoor pours, the evaporation rate depends on weather and is almost wholly beyond the builder's control. On indoor pours, either more heat or better ventilation, or both, will help get rid of bleed water. No one installs heaters just to reduce problems from bleeding, but if the building is heated anyway, it is easy (if costly) to turn up the thermostat. Ventilation may be easier to arrange than heating, but it has little effect if the humidity is very high.

A very effective way to reduce bleeding is to change the concrete mix. In general, any change that increases the amount of very fine particles will help, and so will some other changes. The following steps tend to reduce bleeding:

- use more fine aggregate, especially at the small end of the size range (smaller than 0.150 mm or US number 100 sieve);
- add fly ash or slag;
- use more cement (but beware of the higher cost);
- use less water (but beware the effect on workability and finishability);
- use entrained air (but beware the increased risk of delamination, especially with dry shake finishes).

Setting time

This is defined as the elapsed time between the introduction of water to the concrete mix and the concrete's setting – that is, its hardening. Setting time affects floor quality because it determines the time available for finishing. When concrete sets too fast, finishers lack the time to do their job right. When concrete sets too slow, overtime costs rise and finishers are tempted to skimp on the final steps.

The rate of setting depends on many factors, but most of all on temperature and admixtures.

One factor that, perhaps surprisingly, does not matter much is the choice of cement. All kinds of cement commonly used for floors (see Chapter 9) set at similar rates, though they differ widely in the rate at which they gain long-term strength. While some cements are called "rapid-hardening", the term does not mean "quick setting", but refers instead to the rate of strength gain after setting.

Temperature

Like most chemical reactions, the hydration of cement goes faster at higher temperatures. Heat shortens the setting time; cold extends it. Extreme heat and extreme cold both cause trouble.

It is the concrete temperature, not the air temperature, that controls setting time. But of course air temperature affects the concrete temperature, and only the most massive floors can long maintain an internal temperature much different from the ambient. Measures to control concrete temperature include those that cool or heat the concrete directly, and those that cool or heat the air. The latter tend to be more effective.

When heat makes setting times too short, consider these steps:

- shade the floor from the sun's rays, either by erecting the building's roof and walls or by putting up a temporary shelter;
- place concrete in the cool of the night;
- keep aggregates cool;
- chill the mixing water;
- add ice to the concrete (but be sure to include it when calculating the total water content);
- chill the concrete with liquid nitrogen.

When cold makes setting times too long, consider these steps:

- heat the building (making sure not to use unvented heaters that raise indoor carbon dioxide levels);
- heat the space under a suspended floor;
- heat the mixing water.

Admixtures

Some admixtures change setting time as their primary effect. Others affect setting time as a side effect.

Accelerators shorten setting time; retarders extend it. Most low-range water reducers have a slight retarding effect, which can be used to advantage in hot weather. But other low-range water reducers are mild accelerators, so it pays to read the manufacturers' information sheets.

An admixture overdose, or the use of two admixtures together, may cause a very long delay in setting. Certain impurities in the mix produce a similar effect. A few kilogrammes of sugar in the mixing drum will completely prevent concrete's setting.

Chapter 9 discusses admixtures in more detail.

Accepting short and long setting times

One way to deal with short or long setting times is to live with them, adjusting working methods to finish the concrete well no matter how long it takes.

Fast setting in hot weather causes little trouble if you have enough workers and tools to finish the concrete in the time available. You might expect concrete contractors in the hot climate of the American Southwest to use large quantities of retarders, but most do not. They just accept a short setting time and make sure they have the staff on site to handle it.

Similarly, slow setting in cold weather causes no harm if workers are prepared to stay late. Concrete contractors in some cold regions work split shifts to limit overtime costs. The main crew starts in the morning and stays for a normal workday. The second crew – sometimes just two or three workers – shows up in the afternoon and stays till the bitter end, which may not arrive till the small hours.

Plastic settlement

This is the sinking of the floor surface caused by a reduction in volume of the fresh concrete. Most concrete loses volume from the time it is mixed, through loss of water and of entrapped air. There may also be some leakage through the forms and, in ground-floor slabs, into the sub-base.

Plastic settlement is distinct from drying shrinkage. Both involve reductions in volume, but settlement is a property of plastic concrete while drying shrinkage is a property of hardened concrete. Drying shrinkage occurs in all directions. Plastic settlement occurs only in the vertical direction, because plastic concrete flows horizontally to fill up the forms.

Plastic settlement appears to be proportional to the slab thickness. Slabs thinner than 150 mm (6 in) seldom settle enough to cause trouble. The biggest problems occur in slabs thicker than 300 mm (12 in), and in slabs cast monolithically with thick supporting beams.

Plastic settlement causes trouble in at least two ways: by causing cracks and by making the floor surface less flat.

Plastic-settlement cracks appear in fresh concrete directly over embedded objects such as reinforcing bars or post-tensioning tendons. They occur because the concrete settles and the embedded objects do not. In some cases the whole reinforcing grid appears as cracks on the floor surface. Plastic-settlement cracks are most likely where reinforcing bars or post-tensioning tendons are large in diameter and close to the surface.

Plastic settlement makes a floor less flat because the settlement is not always uniform. If the whole floor surface sank by the same amount, there would be no affect on flatness. But settlement varies from batch to batch and even within each batch, leaving bumps and dips in the surface. Because of plastic settlement, it is very hard to achieve close surface tolerances on thick slabs. If slab thickness exceeds 400 mm (16 in) and the floor needs a superflat or near-superflat finish, the best solution may be a floor topping.

A different flatness problem occurs at slab edges. As the surface settles, the finishers often build up the edge with mortar to keep it flush with the side form. The result is a ridge along the edge. This mimics curling (see Chapter 14) and is sometimes mistaken for it.

Where plastic settlement causes or is likely to cause trouble, one or more of the following steps may be worth a try:

- Use the largest possible coarse aggregate.
- Ensure the coarse aggregate is evenly graded.
- Use less water in the concrete mix (but beware the effect on workability and finishability).
- Leave a generous surcharge when striking off.
- Place deep sections in two or more lifts. This is worth considering when a slab is being cast monolithically with thick joists or beams.
- Add plastic fibres.

8

Properties of Hardened Concrete

This chapter deals with those properties of hardened concrete that affect floors. Freshly mixed concrete sets in a few hours, but important changes continue for a long time after that. These are the properties of hardened concrete that matter most in floors:

- strength;
- impact resistance;
- modulus of elasticity;
- drying shrinkage;
- thermal coefficient.

Strength

Because concrete is a structural material, its strength matters. But because the concrete in a floor is more than just a structural material, we should not consider its strength in isolation from its other properties.

Strength can be defined as the stress needed to make concrete break in a particular way. The basic SI unit for stress, and thus for concrete strength, is the pascal. Concrete strengths fall in the millions of pascals, so the normal unit for concrete strength is the megapascal, abbreviated MPa. The name has never caught on in the British construction industry, which uses megapascals but insists on calling them newtons per square

Table 8.1 Standard tests for concrete strength

Kind of Strength	Standard Test Method	
	British	American
Compressive (cube)	BS EN 12390-4	none
Compressive (cylinder)	BS EN 12390-4	ASTM C 39
Flexural	BS EN 12390-5	ASTM C 78
Tensile (split cylinder)	BS EN 12390-6	ASTM C 496

millimetre (N/mm^2). Americans describe strength in pounds per square inch, abbreviated psi (1 MPa = $1 N/mm^2$ = 145 psi).

Floor designers pay attention to these kinds of concrete strength (Table 8.1):

- compressive strength;
- flexural strength;
- tensile strength;
- shear strength.

Designers who work with steel fibres sometimes uses a strength-related property called ductility or toughness.

Compressive strength

The commonest, though not always the most useful, gauge of concrete quality is compressive strength – resistance to crushing. The UK standard test (found in BS EN 12390) measures the force needed to crush a 150 mm (6 in) cube of hardened concrete. Americans, along with most of continental Europe, replace the cube with a cylinder 150 mm (6 in) in diameter and 300 mm (12 in) high. The American standard is ASTM C 39; BS EN 12390 also includes a test method for cylinders. Because a tall cylinder offers less resistance to crushing than a cube of the same material, American compressive-strength test results are lower than British. The exact relationship between cylinder and cube strengths is complex, but in general cylinder breaks are about 20% lower (Figure 8.1).

In this book I take a divided approach to reporting compressive strength. Strengths that appear in MPa are British cube strengths. Strengths in psi are American cylinder strengths. I apologize to readers in continental Europe, Australia and elsewhere who are used to cylinder strengths

Figure 8.1 Concrete cylinders are used to determine compressive strength in many parts of the world. The UK uses cubes instead

measured in MPa; you should divide the American cylinder results by 145, or reduce the British cube strengths by about 20%.

The arrival of European standards has complicated matters in the UK. British testing laboratories will keep on breaking cubes for the foreseeable future, but the European structural code is based on cylinder strengths. The compromise is to define concrete strength by both methods. British practice divides concrete mixes into strength classes, each bearing two numbers separated by a solidus (Table 8.2). The lower number represents the cylinder strength, in MPa. The higher number represents the cube strength. The letter C tells us this concrete is made with normal or heavyweight aggregates. (The other choice is LC, for lightweight concrete.)

Most concrete for floors has a 28-day compressive strength of 20–40 MPa (2500–5000 psi). Strengths above 40 MPa (5000 psi) are seldom used in full-thickness slabs, because of high cost and drying shrinkage. Mixes for monolithic and bonded toppings sometimes show compressive strength of 50 MPa (6000 psi) or more, however.

Though important and easy to test, compressive strength is overused as a measure of concrete quality. In everyday speech at the building site,

Table 8.2 Compressive-strength classes for normal-weight concrete

	Characteristic Strength		
	150×150-mm Cube	150×300-mm Cylinder	
Class	MPa	MPa	psi
C8/10	10	8	1200
C12/15	15	12	1700
C16/20	20	16	2300
C20/25	25	20	2900
C25/30	30	25	3600
C28/35	35	28	4100
C30/37	37	30	4400
C32/40	40	32	4600
C35/45	45	35	5100
C40/50	50	40	5800
C45/55	55	45	6500
C50/60	60	50	7200

people often speak of compressive strength (sometimes called concrete grade or concrete quality) as if it were the only important thing to know about a mix. But concrete has many other properties that matter in floor construction. Some of those properties have no connection with compressive strength, while others – most notably drying shrinkage – grow worse with higher compressive strength.

Within the field of floor design, compressive strength matters most in suspended slabs. The principles of suspended-slab design call for the concrete to resist compression, relying on steel reinforcement or prestressing tendons to resist tension. Compressive strength is thus essential.

It plays a much smaller role in ground-supported slabs, which normally rely on the concrete's flexural strength to resist bending. The concrete in a ground-supported floor also resists compression, but flexural strength controls the design because concrete's flexural strength is always much lower than its compressive strength.

Compressive strength often stands in for other properties such as flexural strength, shear strength, modulus of elasticity and even wear resistance. Such substitution sometimes makes sense, given that compressive strength is easy to test. But the relationship between compressive strength and other properties varies greatly. If a property truly matters, it usually deserves to be specified and tested directly.

Flexural strength

This is concrete's ability to resist bending. Some people call it "tensile strength in flexure" because while bending creates both compressive and tensile stresses, it is always the latter that cause the concrete to fail.

The usual test for flexural strength measures the force needed to break a beam under third-point loading (see Figure 8.2). Standard test methods appear in ASTM C 78 and BS EN 12390-5. The standard beam is a square prism 150 mm by 150 mm × at least 500 mm (6 in by 6 in by at least 20 in), but other sizes are sometimes used. Flexural strength is expressed as the modulus of rupture, taken from the following equation:

$$f_r = Pl/bd^2$$

where:

f_r = modulus of rupture, in MPa (or psi);
P = force needed to break beam, in N (or lb);
l = length of span (not length of beam), in mm (or in);
b = mean width of beam, in mm (or in);
d = mean depth of beam, in mm (or in).

This equation applies only if the beam breaks within the middle third of the span, as it usually does.

Another test method, called centre-point loading and described in ASTM C 293, breaks the beam by applying force only at mid-span. It gives strength results about 15% higher than those from third-point loading. Since the standard design methods are based on third-point loading, test results from centre-point loading should be reduced before using them in design calculations.

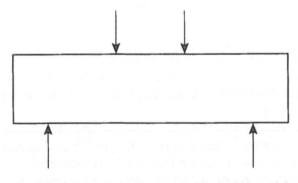

Figure 8.2 To test concrete's flexural strength (also called modulus of rupture), a test beam is broken under third-point loading

Flexural strengths for concrete used in floors ranges from 3 to 5 MPa (400 to 700 psi).

Concrete flexural strength is essential in ground-supported slabs. Unlike suspended slabs that are always designed as reinforced or prestressed members, many ground-supported slabs rely on the concrete alone to support loads. (Ground-supported floor often contain reinforcement, but the steel is meant to control cracks, not to resist applied loads.)

Despite the importance of flexural strength, most floor designers do not specify it directly. They specify compressive strength instead and assume a relationship between the two kinds of strengths. Designers in the UK often assume the following relationship:

$$f_r = 0.393 \, (f_{cu}^{2/3})$$

where:

f_r = modulus of rupture, in MPa
f_{cu} = cube compressive strength, in MPa

American designers often rely on the following equation:

$$f_r = 7.5 \, (f_c^{0.5})$$

where:

f_r = modulus of rupture, in psi
f_c = cylinder compressive strength, in psi

Figure 8.3 shows the approximate correlation between compressive strength and flexural strength. The numbers running up the left side of the graph represent UK-style cube strengths, while those on the right stand for American-style cylinder strengths.

Unfortunately, the British and American equations, as well as the graph in Figure 8.3, imply a precision that does not exist. There is no fixed relationship between compressive and flexural strengths. For any given compressive strength, flexural strength varies widely. It will usually be higher than the standard formulae suggest, but you cannot count on that unless you test it directly.

Where flexural strength is critical – as in heavily-loaded ground-supported slabs – designers should specify it directly and enforce it with beam tests. This is more accurate and, in the long run, more economical than using an assumed relationship between compressive and flexural strength.

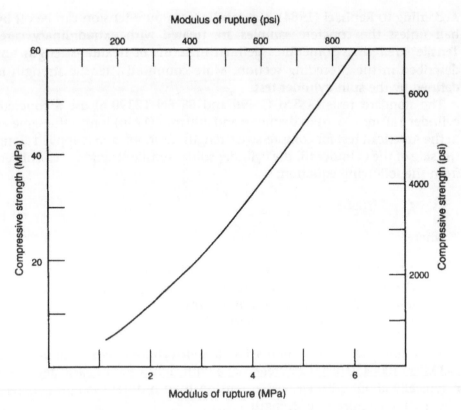

Figure 8.3 Approximation correlation between compressive and flexural strengths

Designers of outdoor pavements have, in many places, accepted this idea. Flexural strength is commonly specified and tested for concrete highways and airport runways. The practice is not yet common in floor construction, but it should be.

Where precise control over flexural strength is unimportant – as in most suspended slabs and in lightly-loaded ground-supported slabs – it makes sense to test for compressive strength and to estimate flexural strength, if necessary, from one of the standard equations.

Tensile strength

This is the concrete's ability to resist being pulled apart. Concrete is only rarely tested in direct axial tension because accurate results are hard to get.

According to Raphael (1984), test results from pure tension can be off by half unless the concrete samples are treated with extraordinary care. Tensile strength is sometimes defined in terms of flexural-strength test described in the preceding section. More commonly, tensile strength is defined by the split-cylinder test.

The standard tests (ASTM C 496 and BS EN 12390-6) use a concrete cylinder 150 mm (6 in) in diameter and 300 mm (12 in) long – the same as in the American test for compressive strength. Compression is applied along the side of the cylinder till the cylinder splits. Tensile strength is calculated from the following equation:

$$f_{st} = 2P/(\pi d l)$$

where:

f_{st} = splitting tensile strength, in MPa (or psi)
P = force needed to split the cylinder, in N (or lb)
d = cylinder diameter, in mm (or in)
l = cylinder length, in mm (or in).

Most concrete used in floors has a split-cylinder tensile strength of 2–3 MPa (300–450 psi). Concrete made with lightweight coarse aggregate is typically about 15% weaker in tension than ordinary-weight concrete of the same compressive strength.

Split-cylinder tensile strength is specified and tested even less often than flexural strength. Most designers rely on an assumed relationship between tensile and compressive strengths, even though this suffers from the same shortcomings as similar assumptions about flexural strength. For years, we assumed tensile strength was proportional to the square root of the compressive strength. Various constants were used, but the following equations are typical:

In SI units:

$$f_{st} = 0.6(f_{cu})^{0.5}$$

where:

f_{st} = splitting tensile strength, in MPa
f_{cu} = cube compressive strength, in MPa.

In fps units:

$$f_{st} = 6.7(f_c)^{0.5}$$

where:

f_{st} = splitting tensile strength, in MPa
f_c = cylinder compressive strength, in MPa

The traditional square-root formulae were never quite right, and were probably adopted because square roots were easy to calculate on a slide rule. Carino and Lew (1982) suggest that tensile strength is proportional to the compressive strength raised to the power of 0.73, at least for ordinary-weight concrete.

But it hardly matters. For almost all purposes related to concrete floors, the standard equations are close enough. And for those rare cases where close control of tensile strength is critical, we can specify and control the property directly.

Shear strength

This is concrete's ability to resist shear failure. A shear failure occurs when concrete cracks in a direction parallel to stress.

Shear strength is not easy to measure, and no good standard test for it exists. Some design methods assume that shear strength is closely related to tensile strength, and that both are proportional to the square root of the compressive strength. The second part of that assumption is probably wrong, but not so far wrong that it causes trouble.

Shear strength matters little in ground-supported slabs. If a ground-supported slab has the flexural strength to support its loads, it is unlikely to fail in shear.

The structural codes do require designers to check shear in suspended slabs. In most cases shear does not control the design. Where it does, the almost universal solution is to add reinforcement or increase slab thickness, not to make the concrete stronger.

Ductility

This applies only to concrete containing steel fibres. Some people call it flexural toughness or equivalent flexural strength.

Ductility, in this context, is defined as the ability to resist bending after the concrete has cracked. Plain concrete has no such ability, but steel-fibre concrete does.

The Japanese have a standard test for ductility, described in JSCE–SF4. (Japan Society of Civil Engineers, 1985) In that test, a beam is stressed under third-point loading, just as in the standard test for flexural strength. But whereas the strength test just measures the load needed to break the beam, the ductility test goes on till the beam has deflected by 3.0 mm. All the while load values are being recorded at close intervals. The test result, called R_{c3}, is the mean load applied throughout the test (up to 3-mm deflection) divided by the load at the moment of cracking. According to the Concrete Society (2003), R_{c3} must be at least 0.3 to be of use in structural calculations.

Two American test methods – ASTM C 1018 and ASTM C 1550 – also measure ductility.

Most design manuals for floors make no mention of ductility and provide no way to take advantage of it. The chief exception is *TR 34* (Concrete Society, 2003), which implies that the normal ground-supported floor will include enough steel fibres to make it ductile and shows how to use R_{c3} in structural calculations. Some steel-fibre manufacturers have developed their own design methods to take advantage of ductility.

While no one doubts that ductility exists, its use in floor design calculations remains controversial.

Characteristic (design) strength

Floor designs are based neither on minimum strength (an almost meaningless term) nor on mean strength (which could seriously mislead), but on characteristic strength. Americans call it design strength.

A concrete's characteristic strength is the value below which only a small percentage of readings can be expected to fall. In the UK, no more than 5% of the test results may fall below the characteristic strength. American rules allow 10%.

Characteristic strength is related to the mean strength as determined by tests, but is always lower than the mean by an amount that depends both on the variability of the test results and on the percentage of readings allowed to fall short. Variability is expressed as the standard deviation. Table 8.3 shows the amounts, expressed as multiples of the standard deviation, by which mean strength must exceed characteristic strength for various allowable percentages of short readings.

For most concrete mixes, the standard deviation ranges from 5 to 12% of the mean strength. A well-controlled mix will have a smaller standard deviation; therefore its characteristic strength will be closer to its mean strength.

Where we lack the test results to determine a meaningful standard deviation, standard and very conservative rules apply.

Table 8.3 Difference between characteristic and mean strength

Maximum Number of Strength Test Readings Allowed to Fall Below Characteristic Value	Amount by Which Mean Strength Must Exceed Characteristic Strength
0.1%	3.00 standard deviations
0.5%	2.58 standard deviations
1%	2.33 standard deviations
5% (UK standard)	1.64 standard deviations
10% (USA standard)	1.28 standard deviations
15%	1.04 standard deviations
20%	0.84 standard deviations

Any strength property for which a standard test exists can have a characteristic value. The concept is not limited to compressive strength, but can also apply to flexural strength, tensile strength and ductility.

Strength gain

Concrete grows stronger with time. The process goes on for decades – perhaps even longer – though at an ever diminishing rate. Concrete gains more than half its ultimate strength in a few weeks.

Concrete mixes all do not gain strength at the same rate. The main factor is the kind of cement. If we take as the norm concrete made with ordinary Portland cement (American Type I or II), then rapid-hardening Portland (American Type III) accelerates strength gain, while low-heat Portland (American Type IV) retards it. Pozzolans such as fly ash and slag also slow the rate of strength gain.

Many admixtures affect the rate of strength gain. Accelerators and superplasticizers tend to produce higher strength at early ages (before 28 days) and lower strength later on. (see Table 9.9 in Chapter 9).

We cannot base floor designs on concrete's ultimate strength, because that strength would not achieved for a long time. It has become standard to control concrete strength at 28 days, but that choice is arbitrary and can result in floors that are overdesigned for the applied loads. Since many concrete floors do not go into use till well after 28 days, it often makes sense to base the structural design on strength after some longer period: 90 days is common, but other periods are used. Some designers call for strength to be tested at the standard 28-day mark, and then assume without further testing that the 90-day strength is 10% higher than the 28-day figure. Assuming a 10% gain from 28 to 90 days is reasonable for concrete made with ordinary Portland cement. However, concrete made with rapid-hardening cement (American Type III) or an accelerating admixture may add less than 10% to its strength from 28 to 90 days.

People often measure concrete strength at three days or seven days, or both – not for contract enforcement, but to determine when to remove formwork or for an early sign as to whether the mix will meet its specified 28-day strength. In post-tensioned floors, concrete strength is sometimes tested at one day or even less to determine when to apply the initial stress.

Drawbacks of higher strength

Is stronger concrete better concrete? Not always. Concrete needs to be strong enough to resist the stresses applied to it, with a reasonable safety factor, but beyond that higher strength can cause more harm than good. The drawbacks of higher concrete strength in floors include monetary cost, increased drying shrinkage, increased curling, loss of load transfer at joints and greater risk of some kinds of cracks.

Stronger concrete usually costs more because it contains more cement – and cement is concrete's most expensive basic component. According to Canadian research into the cost of reinforced, one-way suspended slabs, the cheapest designs are those with the weakest concrete and the strongest reinforcing steel. (Salinas,1980) The savings produced by using weaker concrete more than offset the cost of the stronger steel.

Stronger concrete shrinks more on drying because it contains more cement. The added cement causes a small increase in shrinkage on its own, but the main effect comes from the extra water that accompanies the extra cement. In considering the effect of that water, we need to look at the total amount of water in the mix, not the water–cement ratio. Stronger concrete often has a lower water–cement ratio than weaker concrete, but it will still shrink more if its total water content is greater.

Other drawbacks of high concrete strength are consequences of the increased drying shrinkage. As concrete shrinks, slabs curl, joints open up and cracks appear.

Impact resistance

This is concrete's ability to withstand sharp blows.

All the strength and ductility tests described above apply loads slowly and steadily. Many real floor loads are also applied in that way, but not all.

Some concrete floors must withstand sudden blows from dropped objects or from certain industrial processes.

Though not in wide use, tests are available to determine concrete's impact resistance of concrete. Schrader (1981) describes a test that uses a 4.5-kg (10-lb) drop hammer. The hammer falls 460 mm (18 in) to strike a steel ball resting on a standard concrete test cylinder of 150 mm (6 in) diameter. The cylinder is struck repeatedly till it comes apart. The operator records the number of blows to the first visible crack, and the number of blows to complete failure.

Impact resistance is not closely related to concrete strength. High compressive strength may actually make concrete resistant to impact.

There are few guidelines available to the designer who wants to specify impact resistance. Till more research is done, the best approach is to test an existing floor and use it as the standard.

Modulus of elasticity

This is a measure of concrete's stiffness, defined as the ratio of stress to strain. Stress is the force we apply to the concrete, divided by the cross-sectional area of the specimen. Strain is the deformation caused by the stress, divided by the length of the specimen. Here is the formula:

$$E = \frac{fL}{x}$$

where:

E = modulus of elasticity, in GPa (or psi)
f = compressive stress on concrete in GPa (or psi)
L = original length of concrete specimen in mm (or in)
x = change in length of concrete in mm (or in)

Most concrete used in floors has an elastic modulus between 25 and 33 GPa (3.6×10^6 and 4.8×10^6 psi).

The modulus is needed for some structural calculations, but designers almost never specify or test it directly. Most structural engineers base their calculations on an assumed rough relationship between elastic modulus and compressive strength.

A related property, Poisson's ratio, is used in some structural calculations. When you squeeze a piece of concrete, it becomes both shorter and wider.

The ratio of widening to shortening is called Poisson's ratio. It can be tested (ASTM C 469 offers a standard method), but most designers and design manuals simply assume it to be about 0.2.

Drying shrinkage

Concrete floors start life saturated with water. If the concrete is well cured, it stays saturated till the end of the curing period. It then starts to dry out; as it dries, it shrinks.

Drying shrinkage has a huge effect on ground-supported slabs, where it causes cracks, curling and loose joints. Cracks form because friction restrains the slab from sliding freely as it shrinks. This restraint sets up tensile stresses in the slab. If the tensile stress exceeds the concrete's rather modest tensile strength, the slab cracks. Curling is caused – at least in part – by differential drying shrinkage from the top of the slab to the bottom. The top dries out first because it is exposed to the air. Joints become loose as drying shrinkage widens them, reducing the effectiveness of aggregate interlock, keyways and joint filler.

Suspended slabs shrink as much as their ground-supported cousins, but the effects are less marked. Most suspended slabs contain enough reinforcement or prestress to handle shrinkage cracks; cracks occur, but steel keeps them narrow. Suspended slabs are less likely to curl because they dry out from both top and bottom. But curling does occur in toppings laid over suspended floors, and in slabs on metal deck.

The amount of drying shrinkage ranges from about 0.04% to over 0.10%. It continues for years, but at an ever-slower rate.

The amount of drying shrinkage plays a big part in determining how well a floor performs. Floors are designed to accommodate drying shrinkage, of course, but they always perform better when that shrinkage is low. A design meant for moderate shrinkage may fail if the shrinkage turns out to be high. If you work in a part of the world where drying shrinkage generally is low, you may learn habits that will lead to disaster if applied in a high-shrinkage area.

Considering how much harm drying shrinkage can do, and how much it can vary, you might expect designers to pay it a great deal of attention. Most do not, however. Few designers specify limits to drying shrinkage or ask for shrinkage test results. In effect, they bet on low shrinkage but do almost nothing to ensure it.

It need not be that way. We can specify drying shrinkage and test concrete for compliance. ASTM C 157 describes a standard compliance test that measures shrinkage over 28 days of air drying. The test actually takes 56 days because the specimen must be wet-cured for 28 days before the drying begins. The test result is the shrinkage (reduction in length) divided by the original length, expressed as a percentage. Most normal concrete mixes produce test results between 0.020 and 0.070%.

Specified limits for 28-day drying shrinkage are typically 0.030, 0.035 or 0.040%. Lower values are appealing, but they create the risk that local suppliers will not be able to meet them.

Some parts of the world are notorious for very high drying shrinkage. In such places, even the typical 0.035% specification may prove out of reach. Where that happens, the only reasonable answer may be to change the floor design so it will work despite shrinkage. Possible changes include closer joint spacing, more reinforcement and better load-transfer devices at joints.

Though we can never eliminate drying shrinkage, the following steps can reduce it:

- Use bigger aggregate.
- Choose well-grade aggregates.
- Avoid aggregates with high drying shrinkage.
- Use less water in the concrete mix (but beware the effect on workability and finishability).
- Use less cement (but beware the effect on strength and finishability).

It is sometimes useful to know how much of the total shrinkage has already occurred. For example, you would not want to install a hard joint sealant in a floor that had only experienced 20% of its ultimate shrinkage. You can use the following equations to estimate the amount of shrinkage as a percentage of total shrinkage. (Branson, 1977) They assume that the concrete is poured at a slump of 100 mm (4 in) or less, and that the slab is no more than 150 mm (6 in) thick.

$$s = 100t/(35 + t) \quad \text{for } H = 40$$
$$s = (1.40 - 0.01H)\,[100t/(35 + t)] \quad \text{for } 40 < H < 80$$
$$s = (3.00 - 0.3H)\,[100t/(35 + t)] \quad \text{for } H > 80$$

where:

s = percentage of total shrinkage
t = days since the end of moist curing
H = relative humidity of air, in %.

Thermal coefficient

Like most materials, concrete expands as it heats up and contracts as it cools down. The rate of expansion and contraction is called the thermal coefficient, and it does not vary much. For concretes made with normal-weight aggregates, the thermal coefficient ranges from 10 to 13×10^{-6} per Celsius degree (5.6 to 7.2×10^{-6} per Fahrenheit degree).[1] Lightweight concrete shrinks and expands slightly less, by about 8×10^{-6} per Celsius degree (4.4×10^{-6} per Fahrenheit degree).

Because we cannot do much to change concrete's thermal coefficient, there is a temptation to ignore it. That can be a mistake, however. Thermal changes have a big effect on concrete floors, especially on ground-supported industrial slabs.

The main issue here is thermal contraction, not expansion. Thermal expansion rarely causes trouble in concrete floors (though it does in some other concrete structures), because only in the rarest circumstances does it make the floor bigger than its original, as-cast size.

In its effect on concrete floors, thermal contraction mimics drying shrinkage. Both involve reductions in volume, and both cause cracks and lead to loss of load transfer at cracks and joints. Many of the problems that afflict concrete floors come from the combined effects of drying shrinkage and thermal contraction. When we see cracks and loose joints in a floor we tend to blame drying shrinkage, but in many cases thermal contraction also plays a big role.

The overall movement caused by thermal change can approach, and in some cases even exceed, that caused by drying shrinkage. A good concrete mix might show drying shrinkage of 0.04%. If that mix has a thermal coefficient of 11×10^{-6} per Celsius degree, then a temperature drop of 37 degrees Celsius (67 degrees Fahrenheit) would make the thermal contraction equal the drying shrinkage.

Even where the thermal movement remains less than drying shrinkage, as it does in heated buildings and tropical climates, thermal contraction can be harder to deal with than drying shrinkage because it never stops. Drying shrinkage is a one-way street. It goes on for a long time but at an ever diminishing rate, and after a few years it has, for all practical purposes, stopped. In contrast, thermal changes go on forever as the floor warms up in the spring and cools down in the autumn. The only exceptions occur in buildings, such as cold stores and library depositories, that stay at a constant temperature.

Lacking a way to reduce concrete's thermal coefficient, how can we deal with its effects? The first step is to make sure the floor design accommodates all expected volume changes – those from temperature changes as well as those from drying shrinkage. Many techniques used against drying shrinkage work equally well against thermal contraction. Those techniques include control joints, reinforcing steel, steel fibres, post-tensioning and shrinkage-compensating concrete.

The second step is to repair joints and cracks in wintertime, when the concrete is at the point of greatest thermal contraction. A joint or crack filled in midwinter will normally hold up better than one filled in midsummer, provided that both repairs take place after substantially all drying shrinkage takes place. In heated buildings, the end of a floor's first winter is a good time to fill joints and cracks. Thermal contraction is near its maximum, and because the concrete has undergone several months of drying in the very low humidity caused by the artificial heat, you can be confident that much of the total drying shrinkage has already occurred.

In freezers and cold-storage buildings, floor joints and cracks should not be filled till the floor has stabilized at its design temperature. That limits the choice of fillers, since some do not harden well or at all at very low temperatures.

9

The Components of Concrete

Ordinary concrete has three essential components:
- cement;
- aggregates (usually divided into coarse and fine);
- water.

Concrete may also contain admixtures, additions and fibres.

Cement

Cement is the glue that holds concrete together. If you tried to make concrete without cement, the result would be a loose collection of wet aggregates.

What we commonly call cement is, strictly speaking, hydraulic cement – a powder that hardens in the presence of water. Several types are used in floors.

Nomenclature

Though cements are similar around the world, they go by different names in the UK and America. What's more, UK nomenclature has completely changed in recent years with the introduction of European standards. The old British Standard 12 on cement has vanished, along with those beautifully clear and simple descriptions of cement types. In the old British system,

Table 9.1 Main European cement types (BS EN 197-1)

Type	Description
CEM I	Straight Portland cement, with no more than 5% other material
CEM II	Portland cement with up to 35% one other constituent
CEM III	Portland cement blended with more than 35% blastfurnace slag
CEM IV	Portland cement blended with natural pozzolan
CEM V	Portland cement blended with blastfurnace slag and natural pozzolan or fly ash

ordinary Portland cement was called, amazingly enough, ordinary Portland cement. Cement that resisted sulphates was called sulphate-resisting Portland cement.

The new system, set out in BS EN 197-1, divides Portland cements into five main categories identified by Roman numerals – CEM I to CEM V (see Table 9.1). CEM I is straight Portland cement. The other four CEM categories describe blended cements containing Portland cement mixed with other materials, mostly pozzolans. All categories except CEM I are subdivided, using a complex notation that can identify the proportion of Portland cement, the other component (other than Portland cement, that is), the strength class, and how fast it gains strength. For example, the identifier CEM II/B-S 42.5 N reveals all the following:

- Roman numeral II tells us this is a blended cement with at least 65% Portland cement and one other ingredient.
- The letter B tells us this cement contains a medium proportion of Portland cement, within the range allowed for Type II. (A or C here would mean higher or lower proportion of Portland, respectively.)
- The letter S tells us the other ingredient is blastfurnace slag. (V here would mean fly ash, and D would mean silica fume.)
- The number 42.5 identifies the strength class.
- The letter N tells us this cement gains strength at the normal rate. (R here would identify a cement producing high early strength.)

Though BS EN 197-1 divides cements into dozens of categories, only a few are actually manufactured in the UK. The British Cement Association (2000) lists just six BS EN 197-1 cements being made, and of those only four make reasonable candidates for floor construction (see Table 9.2).

Just to keep things interesting, American standards also divide Portland cement into five categories identified by the Roman numerals I–V.

Table 9.2 BS EN 197-1 cements made in the UK

Type	Description	Use
CEM I	Ordinary Portland cement	Most floors
CEM II/A-LL	Portland cement with 6–20% ground limestone	Not for floors
CEM II/B-S	Slag cement, 65–70% Portland cement with 21–35% blastfurnace slag	Most floors, with special value in hot weather and thick slabs
CEM II/B-V	Fly-ash cement, 65–79% Portland cement with 21–35% fly ash	Most floors, with special value in hot weather and thick slabs
CEM III/A	Slag cement, 35–64% Portland cement with 36–65% blastfurnace slag	Most floors, with special value in hot weather and thick slabs
CEM III/B	Slag cement, 20–34% Portland cement with 66–80% blastfurnace slag	Not for floors

The definitions are completely different, however, except for Type I, which includes ordinary Portland cement in both systems. Table 9.3 lists the American types.

Canadians use the same categories as Americans, but change the Roman numerals to Arabic and multiply by 10. Thus American Type I becomes Canadian Type 10; American Type II becomes Canadian Type 20; and so on. Do not ask why.

Table 9.3 American Portland cements (ASTM C 150)

Type	Description	Use
I	Ordinary Portland cement	Most floors
II	Ordinary Portland cement, with slightly lower heat of hydration than Type I	Most floors, with special value where moderate resistance to sulphate is needed
III	High-early Portland cement	Floor repairs and tight construction schedules
IV	Low-heat Portland cement	Very thick slabs
V	Sulphate-resisting Portland cement	Floors in contact with sulphate-rich ground water

Portland cement

The great majority of concrete floors contain Portland cement, either on its own or blended with a pozzolan. The British standard for Portland cements, including blended cements, is BS EN 197-1. Americans have ASTM C 150 for Portland cement and ASTM C 595 for Portland-pozzolan blends.

Portland cement is made by heating a mixture of limestone and clay to partial melting at about 1400°C (2600°F). After it cools, the material is ground with gypsum into a fine powder. By changing the proportions of the raw materials, and by adjusting the manufacturing process, different kinds of Portland cement are made. Varieties include:

- ordinary Portland cement;
- high-early-strength Portland cement;
- low-heat Portland cement;
- sulphate-resisting Portland cement;
- white Portland cement.

Ordinary Portland cement – CEM I in the UK, or Type I in America – is a general-purpose product suitable for most floors under most conditions. Americans have the added choice of Type II, very similar to Type I but with a slightly lower heat of hydration and better resistance to sulfates. Actually Americans do not always have that choice, because in many part of the USA only one type, I or II, is available. The specifications for Types I and II overlap, with the result that some manufacturers supply a product called Type I-II.

High-early-strength cement is used where the concrete must achieve high strength at an early age. This is what British standards used to call rapid-hardening cement. Now BS EN 197-1 identifies it by the letter R at the end of the name. Americans call it Type III. High-early cement does not set much faster than ordinary Portland, but the hardened concrete gains strength rapidly over the next few days. Rarely needed in ground-supported floors, this cement finds occasional use in multi-storey construction when a fast schedule calls for forms and shores to be removed as soon as possible. It is also used in floor repairs. It has drawbacks: an increased risk of thermal-contraction cracks and higher drying shrinkage. Alternatives to high-early cement include adding an accelerator, or just using more ordinary Portland cement.

Low-heat Portland cement (American Type IV cement) gains strength more slowly and generates less heat than other types. Mainly used to limit heat build-up in massive structures like dams, it is seldom needed in floor slabs, which are usually thin enough that damagingly high temperatures do not develop. If you can find it, it may be worth considering for very thick slabs – 400 mm (16 in) or more – especially in hot weather. An alternative to low-heat cement is to combine ordinary cement with a pozzolan such as fly ash or slag.

Sulphate-resisting Portland cement (American Type V) is sometimes used in ground-floor or basement slabs that come into contact with sulfate-bearing ground water, which can attack ordinary cement. American Type II cement also resists sulfate, though not as well as Type V.

White Portland cement closely resembles ordinary Portland in performance and composition, but it contains almost none of the iron compounds that give other cements their characteristic grey hue. Though the cement is indeed white, floors made with it may not be. The finished product tends toward light grey, influenced by the colour of the fine aggregate and the amount of trowelling. White cement has been used to make light-coloured floors, but nowadays a light-reflective dry shake is more common for that purpose.

Expansive cement

This contains a chemical that expands when mixed with water. It is used to make shrinkage-compensating concrete (see Chapter 16), which relies on a controlled, small expansion of the concrete soon after hardening.

Rare in the UK, expansive cement finds wider use in America, though it is hardly common even there. An American standard, ASTM C 845, lists three varieties, Types E-1(K), E-1(M) and E-1(S) similar in performance but differing in chemistry. Only Type E-1(K) is much used. Most people still call it Type K.

An alternative to expansive cement is to blend ordinary Portland cement with an expansive addition or admixture.

Pozzolans

These are cement-like powders that are blended with Portland cement. Pozzolans are not cements and have little or no cementing ability on their own. But they form cementitious compounds when mixed with Portland cement and water. Pozzolans are sometimes called cement-replacement materials because they can replace some of (never all) the cement in a concrete mix. The combination of Portland cement and pozzolan is called blended cement.

The original pozzolan was volcanic ash. Today three pozzolans are widely used in concrete floors, and all are industrial by-products. They are:

- slag (also called ground granulated blastfurnace slag or ggbs), which comes from steel manufacturing;
- fly ash (also called pulverized fuel ash or pfa), which is extracted from the smokestacks of coal-fired power stations;
- microsilica (also called silica fume), which comes from the silicon industry.

The main reason for using slag or fly ash is to save money. Both products cost less than Portland cement. That's not true of microsilica, however. Microsilica costs more than Portland and is used only where its special benefits justify the cost. Floors made with microsilica, if finished and cured well, have an unusually dense, impervious surface.

Concrete containing a pozzolan differs from a straight-Portland mix in these ways:

- less bleeding;
- lower heat of hydration;
- slower strength gain during its early life.

Those effects are sometimes wanted, and sometimes not. The reduced bleeding is usually beneficial – but not always, since a little bleeding is sometimes needed to keep the floor surface moist throughout the finishing process. The lower heat is beneficial in hot weather but not in cold, where it can result in very long finishing times.

The reduced early strength is the effect that demands the most attention. All concretes made with hydraulic cement grow stronger over time, but not at the same rate. Suppose two concrete mixes – one with pozzolan, and one without – are designed to achieve the same strength at 28 days. Compared to the straight-Portland mix, the pozzolanic concrete will have substantially less strength at 7 days, slightly less strength at 14 days, the same strength at 28 days and increasingly higher strengths at ages beyond 28 days. The reduced early strength need not concern the structural engineer calculating a floor's long-term capacity, but it limits our ability to load the slab during its first few weeks. It is of particular concern on multi-storey buildings where the schedule demands that forms and shores be removed as soon as possible. It also affects post-tensioned floors, where we have to wait longer before stretching the tendons.

The reduced early strength affects curing, too, and the reason for that needs to be explained. Curing can be seen as a race between strength gain and drying. We want the concrete to grow strong enough before drying stops the chemical processes that make concrete stronger. Good curing handicaps the race in favour of strength gain, by slowing the rate at which concrete dries. Pozzolans work in the other direction, by slowing the rate at which concrete gains strength. For that reason, concrete floors made with pozzolans benefit from the best possible curing, and possibly from curing periods longer than normal. Chaplin (1990) suggests that pozzolanic concrete can match ordinary concrete in wear resistance, provided it is wet-cured for at least 7 days or cured with a high-solids resin compound.

In their effect on heat of hydration and strength gain, pozzolans closely resembles low-heat Portland cement (American Type IV), and offer a good alternative to that hard-to-find material.

How much pozzolan can we safely use in floor construction? Opinions differ on that. One study (Lotha, Nautiyal and Jain 1976) on the effect of fly ash showed an increase in concrete creep when that particular pozzolan exceeded 15% of the total cementitious content. Chaplin (1986) reported that quite high levels of pozzolans – up to 35% fly ash and up to 50% slag – did not reduce a floor's wear resistance, provided the concrete was cured well. The Concrete Society (2003, p. 74) recommends limiting fly ash to 30% and slag to 50%.

Non-hydraulic cements

Most concrete floors are made with hydraulic cement – either Portland cement or a near relative, sometimes blended with a pozzolan. When people use the word *cement* with no modifier, they almost always mean hydraulic cement. Non-hydraulic cements exist, however, and are used for some special purposes. Three examples are epoxy, sulphur and bitumen (asphalt).

Epoxy cement makes a very strong concrete with extraordinary resistance to chemical attack. No one makes whole slabs of epoxy concrete – they would cost too much – but it finds use in floor toppings and repairs. Epoxy concrete is hard to finish because the epoxy resin sticks to almost everything, including tools.

Sulphur works as a cement if heated to its melting point – 113°C (235°F). Unlike Portland cement and epoxy, sulphur hardens not by chemical reaction but simply by cooling. The finished product resembles Portland-cement concrete in strength, but has better resistance to some chemicals. It stands up well to acids and most salts, but not to alkalis. It can be finished with the same tools used on Portland-cement concrete, including power trowels.

Bitumen, a thick petroleum product also called asphalt, is the cement in bituminous concrete, used all over the world to pave roads and car parks. It is also used indoors for ground-supported industrial floors and floor toppings, though this is less common than it used to be. Because bituminous concrete flows under sustained heavy loads, it may not be a good choice for floors that support fixed racks or conveyors. The few builders who lay floors of bituminous concrete typically use road-paving machines, including heavy rollers. However, some special formulations are finished more like Portland-cement concrete, using power trowels.

Aggregates

Aggregates are the rock particles that make up the main part of almost every concrete mix. If you tried to make concrete without aggregates, you would get cement paste – a strong material, but one with a high price, very high shrinkage and a propensity to crack.

In the UK, BS EN 12620 is replacing BS 882 as the main standard for concrete aggregates. The American equivalent is ASTM C 33.

Aggregates are inert; that is, they do not enter into the chemical reactions that harden concrete. That does not mean, however, that they are unimportant. They play a large role in determining concrete's workability, finishability, plastic shrinkage and drying shrinkage. They also affect load transfer at cracks and induced joints.

Convention divides aggregates into coarse and fine. Coarse aggregate, also called stone or gravel, consists of particles at least 5 mm (3/16 in) across. Fine aggregate, also called sand or fines, consists of particles less than 5 mm (3/16 in) across.

Though useful, the distinction between coarse and fine aggregates is arbitrary. While particles of different size affect concrete's behavior in different ways, there is no special dividing line at the 5 mm (3/16 in) mark. The conventional distinction between coarse and fine aggregates makes sense mainly because the two usually come from different pits and are stored separately. A few lucky concrete plants lie near a source of sandy gravel that provides all the particle sizes needed for good concrete, and in those cases the distinction between coarse and fine aggregates ceases to exist. Even where coarse and fine aggregates come separately, it makes sense to consider them together when analyzing the gradation.

Coarse aggregate

This consists of rock particles at least 5 mm (3/16 in) across, and it forms the biggest part of most concrete mixes. Its role is to provide bulk, reduce drying shrinkage and help transfer load across cracks and induced joints.

If you tried to make concrete without coarse aggregate, you would get grout or mortar – a material that would cost too much and shrink too much to use in floors.

Coarse aggregate may consist of natural gravel or man-made crushed rock. Both can do the job, but there are some difference between them. Gravel particles tend to be smooth and rounded, making concrete that is easy to pump and strike off. Crushed-rock particles tend to be rougher and more angular, making concrete stronger and improving aggregate interlock at cracks and induced joints.

Coarse aggregate plays a vital role in controlling concrete shrinkage, which is of great concern in floor construction. It also helps transfer loads across cracks and induced joints, a factor of great importance in unreinforced and lightly reinforced floor slabs. Because coarse aggregate is almost always the most voluminous component of a concrete mix, its weight has a big effect on the weight of the finished product. When a design calls for lightweight or heavyweight concrete, the goal is achieved by choosing unusually light or heavy coarse aggregates.

Coarse aggregate has little to do with the properties of the floor surface because it rarely appears there. Floor finishing pushes the bigger particles down, leaving at the surface a paste made of cement, fine aggregate and water.

Gradation of coarse aggregate

The coarse aggregates used in most concrete floors contain a variety of particle sizes. The bigger particles keep the concrete dimensionally stable, reducing both plastic and drying shrinkage. The smaller particles fill the gaps between the bigger particles. Without those smaller particles, the mix would need more fine aggregate and more cement, and would cost more.

Gradation is the division of aggregate into categories defined by particle size. The usual method is sieve analysis, in which the aggregate passes through a set of standard sieves, each finer than the last. The amount of material retained on each sieve is weighed. The weights are then used to calculate the percentage of material that passes each standard sieve.

Tables 9.4 and 9.5 show the gradation limits for coarse aggregates in the UK and America. UK sieve sizes are straightforward. American sizes are more complicated. The bigger sizes, like their UK counterparts, are identified by a dimension that shows the width of each square hole in the sieve. That is simple enough, but sizes below 3/8 in are identified by numbers that bear no clear relationship to any dimension in millimetres or inches. Higher numbers denote smaller sieves.

Table 9.4 British gradation limits for coarse aggregate (BS 882)

Nominal Size	Amount Passing Each Sieve (% Mass)					
	37.5 mm	20.0 mm	14.0 mm	10.0 mm	5.0 mm	2.36 mm
Graded 40–5 mm	90–100	35–70	*	10–40	0–5	*
Graded 20–5 mm	100	90–100	*	30–60	0–10	*
Graded 14–5 mm	100	100	90–100	50–85	0–10	*
Single-size 40 mm	85–100	0–25	*	0–5	*	*
Single-size 20 mm	100	85–100	*	0–25	0–5	*
Single-size 14 mm	100	100	85–100	0–50	0–10	*
Single-size 10 mm	100	100	100	85–100	0–25	0–5

*not specifically controlled in BS 882

Making the situation even more complex, Americans use another system to define aggregate gradation. ASTM C 33 identifies coarse aggregates by "size numbers" that range from 1 to 8. In this system, a single batch of aggregate may bear two or three numbers if it contains a mix of sizes. For example, the widely used number 57 aggregate includes sizes 5 and 7, while number 467 includes sizes 4, 6 and 7. (They should be pronounced "five seven" and "four six seven", by the way.) Table 9.5 lists the gradations for the ASTM C 33 size numbers likely to be encountered in floor construction. The most common size numbers are 467, 57 and 67 (see Table 9.6 for British equivalents).

Table 9.5 American gradation limits for coarse aggregate (ASTM C 33)

Class	Amount Passing Each Sieve (% Mass)						
	37.5 mm (1½ in)	25.0 mm (1 in)	19.0 mm (3/4 in)	12.5 mm (1/2 in)	9.5 mm (3/8 in)	4.75 mm (No.4)	2.36 mm (No.8)
4	90–100	20–55	0–15	*	0–5	*	*
467	95–100	*	35–70	*	10–30	0–5	*
5	100	90–100	20–55	0–10	0–5	*	*
56	100	90–100	40–85	10–40	0–15	0–5	*
57	100	95–100	*	25–60	*	0–10	0–5
6	100	100	90–100	20–55	0–15	0–5	*
67	100	100	90–100	*	20–55	0–10	0–5
7	100	100	100	90–100	40–70	0–15	0–5
8	100	100	100	100	85–100	10–30	0–10

*not specifically controlled in ASTM C 33
Note: ASTM C 33 includes more gradation classes than shown here. There are the classes likely to be encountered in floor construction.

Table 9.6 Graded coarse aggregates – UK and American equivalents

British (BS 882) BS 882 Nominal Size	Nearest American Equivalent ASTM C 33 Size Number
Graded 40–5 mm	467
Graded 20–5 mm	57
Graded 14–5 mm	67

Most authorities favour a broad, even gradation, with a wide variety of particle sizes and with similar amounts retained on each standard sieve (within limits). Advocates argue that such a gradation produces a highly finishable mix with low drying shrinkage. In America some floor specifications require that coarse aggregate meet the "8-18" rule – at least 8% and no more than 18% retained on each standard sieve between 19.0 mm (3/4 in) and 4.75 mm (Number 4). That is stricter than the normal limits imposed by ASTM C 33.

Not everyone agrees with all that, however. Ian Burnett, an Australian master of concrete mix design, has years of experience making good concrete with just two single-size aggregates: 20 mm and 10 mm (3/4 in and 3/8 in). And single-size coarse aggregate, typically 10 mm (3/8 in), finds use in thin toppings.

Maximum particle size

Whether the gradation is even or not, it has to top out somewhere. Ordinary structural concrete typically contains nothing bigger than 20 mm (3/4 in), or at most 25 mm (1 in). Although such concrete is widely used in floors, there are good arguments for going bigger. Adding bigger stones offers these benefits:

- reduced drying shrinkage, which in turn means less curling, fewer cracks and more stable joints;
- better aggregate interlock at cracks and induced joints;
- less plastic settlement (most valuable in thick slabs);
- a reduction in the amount of cement needed to achieve the same concrete strength.

Drawbacks include:

- difficulties in pumping;
- problems placing and compacting concrete around dense reinforcement;

- cost. (In some markets bigger stones are only available on special order, which drives up the price.)

On most projects the benefits outweigh the drawbacks, at least in ground-supported floors. It usually makes sense to use the biggest aggregate available, subject to these limits:

- The size must not exceed one half the minimum distance between reinforcing bars or wires.
- The size must not exceed one half (some say one third) the slab or topping thickness.

Lightweight coarse aggregate

Ordinary coarse aggregate weighs about 2400 kg/m^3 (150 lb/ft^3). Lighter materials are available, however. Coarse aggregates weighing less than 880 kg/m^3 (55 lb/ft^3) are called lightweight. Lightweight aggregates include pumice (a volcanic rock), as well as artificial products.

Because coarse aggregate is the most plentiful component of the concrete mix, a change to lightweight coarse aggregate makes concrete much lighter. Concrete with ordinary aggregates weighs about 2400 kg/m^3 (150 lb/ft^3). If we replace the coarse aggregate with lightweight material, the weight of the finished concrete falls to about 1700 kg/m^3 (105 lb/ft^3) – a 29% reduction. Even lower weights, down to about 960 kg/m^3 (60 lb/ft^3), are possible if we replace all the aggregates – coarse and fine – with lightweight materials, but that further change makes concrete hard to finish and is rarely done in floor construction.

Almost never used in ground-supported floors, lightweight aggregate offers at least two benefits in suspended floors. It improves the floor's structural capacity by reducing dead weight, and it makes the floor more resistant to fire damage.

The main drawbacks to using lightweight concrete in floors are cost and finishing difficulties. Lightweight concrete can cost half as much again as ordinary-weight concrete.

Deleterious ingredients

The perfect coarse aggregate would consist of nothing but sound, solid particles of rock. Real-world aggregates all fall short of that ideal – sometimes far short. They contain pieces of wood and rubber, lumps of clay, chunks of coal and other materials that do concrete no good.

Those deleterious materials rarely appear at the floor surface, because the finishing process covers them with cement paste. They sometimes end up just below the surface, however, and that is where they cause trouble. The thin layer of cement paste above them may break up under traffic, leaving a hole. Or worse, the offending particle may expand from absorbing water or from chemical reaction, bursting the surface and creating a popout.

Standards limit the amount of deleterious material in coarse aggregate, but the amounts allowed are greater than you might suppose. ASTM C 33 allows "clay lumps and friable particles" to make up 5% of the total mass. You probably would not buy a bag of rice or a box of cornflakes knowing that up to one twentieth of it might be inedible. But when you buy coarse aggregate specified to meet ASTM C 33, you might have to accept that much undesirable material. (The standard actually allows up to 6.5% deleterious ingredients, if we add up all possible kinds.)

Of course, not every coarse aggregate tests the limits. But all contains some impurities, and for that reason we should never be surprised when a concrete floor develops a few small holes or popouts.

Alkali-aggregate reaction

Certain aggregates react chemically with the alkalis present in Portland cement. They expand over time, causing popouts.

The obvious solution is to avoid such aggregates. But that's not always practical, because in some regions all available aggregates are reactive. Builders in those regions have learned ways to reduce the risk. The trick is to minimize the amount of time the aggregates remain damp, because the damaging chemical reaction occurs only in the presence of water. Techniques include pouring indoors so the slab never receives rain, and using a curing compound instead of a wet cure.

ASTM C 33 says the use of low-alkali cement or the addition of pozzolans can reduce the damaging expansion from alkali-aggregate reaction.

Fine aggregate

This consists of rock particles less than 5 mm (3/16 in) across. Its role is to fill the voids between the pieces of large aggregate. It also forms part of

Table 9.7 British gradation limits for fine aggregate (BS 882)

Sieve	Amount Passing (by Mass)		
	Grade C	Grade M	Grade F
10.0 mm	100%	100%	100%
5.00 mm	89–100%	89–100%	89–100%
2.36 mm	60–100%	65–100%	80–100%
1.18 mm	30–90%	45–100%	70–100%
0.60 mm	15–54%	25–80%	55–100%
0.30 mm	5–40%	5–48%	5–70%
0.15 mm	0–15%	0–15%	0–15%

the mortar paste which makes a concrete floor finishable. You can make concrete without fine aggregate, but the result – called no-fines concrete – is almost never used in floors.

Gradation of fine aggregate

Fine aggregates are graded by sieve analysis in much the same way as coarse aggregates, but the results are interpreted differently.

British practice (see Table 9.7) divides fine aggregates into three gradation categories called grades C, M and F. While all three grades have been used successfully in floors, Chaplin (1990) recommends only grades M and F for floors that need high resistance to wear. The Concrete Society (2003) warns that we should not build industrial floors with fine aggregate "at the coarse end of grade C or the fine end of grade F".

American practice (see Table 9.8) does not divide fine aggregates into categories. One gradation table (ASTM C 33, paragraph 6.1) applies to all.

Table 9.8 American gradation limits for fine aggregate (ASTM C 33)

Sieve	Amount Passing (by Mass)
9.5 mm (3/8 in)	100%
4.75 mm (No. 4)	95–100%
2.36 mm (No. 8)	80–100%
1.18 mm (No. 16)	50–85%
0.60 mm (No. 30)	25–60%
0.30 mm (No. 50)	5–30%
0.15 mm (No. 100)	0–10%

Americans do, however, add a refinement called the fineness modulus. To calculate it from sieve analysis, add the percentages retained on each of the following sieves: 0.15 mm (Number 100), 0.30 mm (Number 50), 0.60 mm (Number 30), 1.18 mm (Number 16), 2.36 mm (Number 8), and 4.75 mm (Number 4). Divide that sum by 100 to get the fineness modulus. To meet ASTM C 33, the fineness modulus must not be lower than 2.30 nor higher than 3.10. A high fineness modulus mean the aggregate is deficient in smaller particles.

Natural sand versus crushed-rock fines

Like coarse aggregates, fine aggregates may be either natural or man-made. In the case of coarse aggregates, neither source is consistently superior. But in the case of fine aggregates, natural sand is almost always the better choice. The man-made product, called manufactured sand or crushed-rock fines, often contains too much dust and has a particle shape that makes it hard to finish. Some designers specify nothing but natural sand, especially for industrial floors. The British standard for concrete wearing surfaces, BS 8204:Part 2, calls for natural sand in floors that need high resistance to wear. On the other hand, the Concrete Society (2003, p. 75) suggests that manufactured sand is safe in industrial floors, as long as no more than 9% of it, by mass, passes a 0.075-mm sieve.

In the face of these conflicting recommendations and the absence of any good research results, should we still prefer natural sand? I say yes, but we should not be dogmatic about it. Where high-quality natural sand is available at a reasonable price, use it. But where it costs far more than manufactured, or is of low quality, consider the man-made alternative.

Water

Water is an essential ingredient of every concrete mix, except those few mixes made with non-hydraulic cements. The properties of the water get little attention because they do not vary much. Almost all drinkable water makes good concrete, and much non-potable water will work as well. Decisions involving water tend to centre on the amount added to the mix.

Concrete (more precisely, hydraulic-cement concrete) needs water for the chemical reactions, called hydration, that make concrete hard and strong. All concrete mixes contain more water than the minimum needed to hydrate the cement. The extra water, sometimes called water of convenience, lubricates the mix, making it easier to place and finish. On the other hand, extra water reduces the concrete's strength and increases its drying shrinkage.

Sea water

In a few parts of the world, scarcity of fresh water demands that concrete be made with sea water or other salt water. The main drawback to that practice is that the finished product will contain more chloride ions than national codes allow for reinforced or prestressed concrete. Salt water is safe to use in plain, unreinforced concrete, however. And it may be usable in reinforced or prestressed concrete if the embedded steel is protected from corrosion.

A less important drawback to sea water is that it causes efflorescence – salt deposits on the floor surface. The problem is mainly one of appearance.

Water temperature

The temperature of the mixing water has a big effect on the temperature of the fresh concrete, because water has a much higher specific heat (heat capacity per unit of mass) than cement or aggregates. When concrete needs to be made cooler or hotter, adjusting the water temperature is generally more effective than trying to change the temperature of the other components. In many parts of the world, ready-mix plants heat their water in winter. Chilled water, though less common, is available in some markets.

Where ice is used to cool concrete, its mass must be counted in the mix's total water content.

Admixtures

Plain concrete consists of cement, aggregates and water. An admixture is any other material, except fibres, deliberately added to the concrete mix for a specific purpose.

You may occasionally see the related term *addition*. Strictly speaking, an addition is added to the dry cement, often during the cement-grinding process. In contrast, an admixture is introduced later when the concrete is batched. Additions are dry. Most admixtures are liquid.

Admixtures are often useful, sometimes essential, and widely overused. Thanks to manufacturers' sales efforts, we hear a lot about the benefits of admixtures. We hear less about their unwanted side-effects. For example, some admixtures increase drying shrinkage. ASTM C 494, "Standard specification for chemical admixtures for concrete", allows such an increase as long as it does not exceed 35% (20% in the case of air entrainers). Given that drying shrinkage causes some of the most troublesome floor problems, you might not wish to specify or buy an admixture that causes even a 5% increase in that property, much less 35%. But if tests showed a product causing, say, a 34% increase, you might never hear the news. You would only be told the product "meets ASTM C 494".

Admixtures always cost money, but at least two kinds – accelerators and low-range water reducers – produce offsetting savings that sometimes reduce their nett cost to zero or less. Accelerators save labour because they shorten setting times. That property is particularly valuable in cold weather. Low-range water reducers save money because they allow a reduction in the amount of cement to achieve a given strength.

Admixtures are classified by their effects, not their chemistry. A category may contain widely different chemicals, but their intended effects will be similar. Some chemicals affect concrete in more than one way, and thus are present in more than one category. The following kinds of admixtures are used in floors:

- air entrainers;
- accelerators;
- retarders;
- low-range water reducers (plasticizers);
- mid-range water reducers;
- high-range water reducers (superplasticizers);
- pigments;
- shrinkage reducers.

Air entrainers

These admixtures create air bubbles in the cement paste. Entrainment is the deliberate creation of tiny voids in the concrete, while entrapment is the incidental creation of much bigger voids during the mixing and placing of the concrete (see Chapter 11). Entrained air bubbles are round and have a diameter of 0.02 to 0.50 mm (0.0008 to 0.02 in). Entrained air is usually expressed as a percentage of the total concrete volume. Typical values range from 3 to 6%.

Air entrainment is usually accomplished through use of an admixture added when the concrete is batched. North Americans, who use a lot of air-entrained concrete, have the alternative of cement that contains an air-entraining addition. Such cement is identified by the letter "A" after the Roman-numeral. For example, Type IIA cement entrains air; Type II does not.

The main reason, and the only really good one, to entrain air is to help hardened concrete resist frost damage. The air spaces provide room into which water can expand as it freezes. For good frost resistance, designers specify entrained-air contents of 4 to 6%, depending on the size of the biggest stones in the mix. Coarser aggregates allow lower air contents.

In addition to its effect on frost resistance, entrained air also makes the mix more workable and reduces bleeding. To achieve those goals on floors where frost resistance is not an issue, designers sometimes specify air contents around 3%.

Entrained air's main drawback – a huge one – is an increased risk of delamination. Finishing tends to trap the air under a thin surface layer, which then fails to bond properly to the main mass of the concrete below. The risk rises with the smoothness of the finish. The use of dry-shake finishes greatly increases the risk, and for that reason the American Concrete Institute recommends against any entrained air on slabs that get such dry shakes.

A lesser drawback is reduced concrete strength. Each percentage point of entrained air lowers the concrete's compressive strength by about 5%.

Most concrete floors need no entrained air to resist frost. Two conditions must be met for frost to damage mature concrete, and few floors meet them. First, the concrete must undergo repeated freezing and thawing. A one-time fall below the freezing point, as occurs in a cold store, rarely does noticeable harm. Second, the concrete must be saturated with water when freezing occurs. Almost no interior floor slabs meet both those conditions after the building is completed. Some floors do meet both conditions during construction, however, and in those cases people argue whether the benefit of entrained air outweighs the risk.

Please note that entrained air does not protect newly placed concrete from frost damage. It works only on mature concrete that already gained substantial strength.

Accelerators

Accelerators make concrete set earlier, and they cause the hardened concrete to gain strength faster. They may slightly reduce long-term strength (see Table 9.9).

Table 9.9 Potential effect of accelerating admixture on compressive strength (from ASTM C 494)

Concrete Age	Minimum Compressive Strength, Compared to Unaccelerated Control Mix
3 days	125%
7 days	100%
28 days	100%
6 months	90%
1 year	90%

ASTM C 494 requires that accelerators produce certain effects relative to an unaccelerated control mix. They must shorten setting time by at least 1 hour and at most $3\frac{1}{2}$ hours. Their effect on compressive strength varies with time (see Table 9.9). At three days, the accelerated mix is supposed to be at least 25% stronger than the control mix. After six months, the accelerated mix can be as much as 10% weaker.

Accelerators are used in cold weather to reduce late-night work and overtime expense. They also allow earlier removal of forms and shores, an important factor on some multi-storey projects.

The traditional accelerator is calcium chloride. It is cheap and effective, but in a damp environment it (or any other chemical that releases chloride ions) promotes corrosion of steel embedded in the concrete. For that reason, structural codes prohibit the use of calcium chloride in reinforced and prestressed (including post-tensioned) concrete.

That might seem the end of the story for calcium chloride, but the chemical still has a place in unreinforced floors. And some designers allow it in reinforced ground-supported slabs, most of which lie beyond the restrictions imposed by the structural codes. The truth is that calcium chloride is not nearly so dangerous in floor slabs as the warnings suggest, because it only promotes corrosion while the concrete is damp. Most concrete floors are dry almost all the time, and in such floors calcium chloride is harmless.

For those who need an accelerator but cannot or will not use calcium chloride, non-chloride accelerators are available at higher cost.

Whatever accelerator is used, the dosage should stay the same throughout each pour. Some builders try to save money by dosing just the last batch or two, but that practice causes the last batches to set before the others, interfering with the orderly sequence of finishing steps.

Retarders

Retarders delay the setting of concrete. They have little effect on strength gain after setting.

Retarders are used mainly in hot weather. High temperatures make concrete set faster, leaving less time for finishing. Retarders give the floor finishers more time to do their job. They are worth considering when air temperatures exceed 30°C (86°F).

Retarders are sometimes used in cooler weather if a long delay between concrete mixing and placing cannot be avoided. That happens where the job site lies far the concrete plant.

Retarders are rarely essential. Many builders working in hot climates have developed other ways to deal with short setting times. They pour at night when the air is cooler, or they just make sure they have enough labour on hand to finish the concrete in the short time available.

Low-range water reducers (plasticizers)

These admixtures increase concrete's workability. They are called water reducers because they reduce the amount of water needed for a workable mix. The "low-range" part of the name distinguishes them from mid-range and high-range water reducers, which do the same thing but in greater measure.

Low-range water reducers let us reduce water content by 5 to 12% while keeping the same workability. That makes concrete stronger and tends to reduce drying shrinkage. Alternatively, though less often, water reducers are used to increase workability while leaving water content unchanged.

Many water reducers have mild retarding effect, which may or may not be wanted.

Though a few people still do not like them, low-range water reducers are as safe as any admixtures can be, and they are very widely used.

Mid-range water reducers

These increase workability, with effects between those of low-range and high-range water reducers. Or to put it more precisely, they produce a range of water reduction that overlaps that of the other two categories. By definition, low-range water reducers let us reduce water by 5–12% while retaining the same workability. High-range water reducers allows a reduction of more than 12% in water content.

Mid-range water reducers cross that 12% dividing line. At low doses they serve as low-range water reducers, and at high doses they serve as high-range water reducers.

Unlike high-range water reducers, mid-range products do not lose effect after 40 minutes or so.

High-range water reducers (superplastizers)

These increase workability, letting us reduce water content by at least 12% while keeping the same workability.

They allow the use of very low water–cement ratios that would otherwise make concrete almost impossible to mix and place. Such mixes have only a small role in floor construction, however, because they lack the free water needed for floating and trowelling.

In floors, high-range water reducers are more often used with normal water–cement ratios to make high-slump, so-called flowing concrete. Slumps of 200 mm (8 in) are possible. Such concrete is easy to place and strike off, but there is a price. Floor flatness often suffers because the concrete changes shape so readily before it has set. Compared to a plain mix, flowing concrete made with a superplasticizer may need about 5% more fine aggregate. (Blackledge, 1980, p. 11)

Some contractors rely on flowing concrete to lay large areas in wide bays, using wet screeds. That technique has dwindled in popularity as powered laser screeds have come into wide use. Laser screeding allows, and indeed works better on, stiffer mixes.

Some concrete finishers dislike high-range water reducers, for at least two reasons. First, they complain of uneven setting in the form of "mushy spots" – small areas that set much slower than the rest of the floor. Poor mixing would seem the likely culprit, but longer mixing times have not always eliminated the problem. Second, they have to deal with the chemical's very short effective life. Unlike low-range and mid-range water reducers, superplastizers switch off suddenly about 40 minutes after batching. The resulting near-instant stiffening of the concrete can interfere with floor finishing.

Pigments

Various pigments are available as admixtures (see Table 9.10). Some pigments are also available as additions, to be blended with the dry cement.

Table 9.10 Pigments

Colour	Pigment
Grey and black	Iron oxide
	Mineral black
	Carbon black
Blue	Ultramarine blue
Red	iron oxide
Brown	iron oxide
	Raw and burnt umber
Ivory and cream	Yellow iron oxide
Green	Chromium oxide
	Phthalocyanine green
White	Titanium dioxide

Pigmented concrete is hard to get right, particularly in floors. The results are often disappointing, for at least three reasons. One problem is that pigments affect only the cement paste, leaving the aggregates to show their natural hues. Because of this, the colours are never very intense. Anyone expecting the colour intensity of, say, injection-moulded plastic is sure to be disappointed.

A second reason for disappointing results is efflorescence. Efflorescence is the deposition of whitish salts on the concrete surface. Pigments do not cause efflorescence, but they make it stand out.

The third problem is the difficulty of getting consistent colour from batch to batch. Small variations, which might be unobjectionable in uncoloured concrete, are painfully obvious on a coloured floor.

If you use pigments despite these problems, always provide a test slab so the user can judge the results. The simple step can prevent many arguments and costly repairs.

Remember, too, that there are other ways to colour a concrete floor. Coloured dry shakes may provide greater intensity, though they still face problems of efflorescence and colour variation. If a user demands a very bright, consistent colour, the best answer is probably a floor coating.

Shrinkage reducers

The late 1990s saw the commercial introduction of admixtures that reduce concrete's drying shrinkage. They work by reducing the surface tension of water.

Since drying shrinkage lies at the root of the most intractable floor problems – cracking, curling and loss of load transfer at joints – the availability of an admixture that substantially reduces drying shrinkage seems too good to be true. Alas, it may be just that. The admixtures are dear, and they work best on mixes with water–cement ratios lower than those normally used in floor construction.

Fibres

These are short, thin filaments added to a concrete mix. Many kinds are available, but almost all fibres used in floors fall into one of two broad categories: steel and synthetic (plastic). The two categories serve completely different purposes.

Steel fibres are used for their effect on hardened concrete. At low and medium dosage rates – from 15 to 40 kg/m^3 (from 25 to 65 lb/yd^3) – they help limit crack width and improve load transfer at cracks and induced joints. At high dosage rates – above 40 kg/m^3 (70 lb/yd^3) – they can prevent visible cracks by stopping the spread of micro-cracks. Though steel fibres have almost no effect on concrete's strength, as measured in the usual ways, they do increase its resistance to bending after it has cracked – a property called ductility or toughness. Some design methods take that into account in determining a slab's structural capacity.

Steel fibres range in length from 12 mm (1/2 in) to 60 mm (2-3/8 in). The longer fibres are good for limiting crack width at low and middle dosage rates. The shorter fibres are better for preventing cracks at high dosage rates, because they can be added in greater numbers. Some fibres have hooked ends or a wavy profile to help them grab the concrete. However, plain straight fibres are also used with success.

In contrast, synthetic fibres are used mainly for their effects on plastic (that is, not yet hardened) concrete. Advocates claim that synthetic fibres reduce segregation, plastic settlement cracks and plastic shrinkage cracks. Most plastic fibres are made of polypropylene or nylon, and are added at about 1 kg/m^3 ($1\frac{1}{2}$ lb/yd^3).

10

Mix Design and Mixing

Mix design is the process of deciding how much of each component goes into each cubic metre or cubic yard of concrete. Mixing is the actual combining of those components to make concrete.

Mix design

In modern construction, the person who designs the floor is rarely the person who designs the concrete mix. Most concrete comes from ready-mix companies, who hire specialists to design their mixes. Nevertheless, floor designers – and builders, too – need to know something about the subject so they can ask the right questions and comment intelligently on mix designs prepared by others.

There are three basic kinds of concrete mixes:

- ratio mixes (also called nominal mixes);
- prescribed mixes and designated mixes;
- designed mixes.

Almost all concrete from ready-mix plants falls in the last category. Ratio mixes and prescribed mixes still appear on small jobs where concrete is mixed on site.

Ratio mixes

Also called nominal mixes, these are described by the ratio of cement to fine aggregate to coarse aggregate, either by volume or weight. The amount of water is not defined, under the assumption that the batcher will add just enough water to attain the desired workability. This used to be the standard way to describe all concrete mixes, but prescribed and designed mixes have superseded it for most purposes.

Cement–sand screeds take a slightly different method because they lack coarse aggregate. Here the specification consists of just two numbers, expressing the ratio of cement to sand. The most common screed mixes are 1:3 and 1:4, by volume.

Prescribed and designated mixes

In the UK, these occupy the middle ground between crude ratio mixes and finely tuned design mixes. Both are standard recipes intended for specific, limited purposes. They avoid the extra work needed to prepare and review a designed mix.

Prescribed mixes, spelled out in BS 5328, are meant for small projects where concrete is mixed on site and no one tests the concrete for strength. You can prepare a prescribed mix in your back garden, if you are so inclined. BS 5328 recommends standard mix ST4, with a target slump of 75 mm (3 in), for house and garage slabs. No other concrete floors are likely to be made with a prescribed mix.

Designated mixes are similar to prescribed mixes, but are meant for ready-mix plants and larger site-mixing operations.

Designed mixes

These are tailor-made for specific requirements. Most are prepared by ready-mix concrete firms, though a few floor designers do the job themselves. Good mix design demands not only an understanding of the principles involved, but also knowledge of how those principles apply to the available materials, which vary greatly from place to place.

If you need to or wish to learn the finer points of mix design, Murdock, Brook and Dewar's *Concrete Materials and Practice* (1991) is a good place to start, and is unusual in that it deals clearly with both British and American procedures. Beyond that, most mix designers use a standard method. In the UK the choices include the BRMCA (for British Ready Mixed Concrete Association), (Dewar, 1986) and the simplified DoE method (Department of the Environment, 1988). In America the most popular method appears in ACI 211, *Standard Practice for Selecting Proportions for Normal, Heavyweight and Mass Concrete.*

With any method, mix design requires a high level of skill. Fortunately it is not a skill floor designers must possess. You can design a concrete floor without knowing how to design a concrete mix (people do it every day), provided you know what questions to ask. The usual approach is for the floor designer to specify certain properties the concrete must meet, and then to let the concrete-mix designer suggest a recipe that will meet those requirements. Floor designers typically specify some of the following properties:

- compressive strength;
- flexural strength;
- drying shrinkage;
- maximum or target slump;
- water–cement ratio;
- entrained air;
- minimum cement content;
- cement type;
- gradation of coarse aggregate;
- allowability of admixtures.

When a designed mix is submitted for review, it always includes a list of ingredients showing the amounts contained in each cubic metre (or cubic yard) of concrete. It often includes test results that show the mix is likely to achieve the specified strength. It may not contain much more unless someone insists, and on important jobs, someone should. To allow proper evaluation, a mix-design submittal may need to include some or all of the following information beyond the basics:

- coarse-aggregate gradation showing the amount retained on each standard sieve;
- fine-aggregate gradation showing the amount retained on each standard sieve;
- combined aggregate gradation showing the amount that would be retained on each standard sieve if both coarse and fine aggregates were sieved together;
- composition of the coarse aggregate – type of rock, and whether natural gravel or crushed rock;
- composition of the fine aggregate – whether natural sand or crushed rock fines;
- test results for drying shrinkage.

Keeping an open mind

Some key concrete properties are hard to predict. Few of us can look at a mix design on paper and know for sure the concrete will be finishable or pumpable.

For that reason, we should keep an open mind about every mix design. If a mix proves hard to finish or turns out to have other bad properties, we should be ready to change it, even if the job schedule does not leave time for full testing of the changed mix design.

When a new mix design is used in a demanding floor, it is wise to start with a trial slab in a non-critical part of the building.

Mixing

We have two choices here: ready-mix or site-mixed concrete.

Ready-mix concrete

This is batched in a concrete plant and delivered to the site in trucks. Ready-mix concrete has become the norm for almost all floor construction, at least in the industrialized world, except the very smallest and very biggest jobs.

Ready-mix plants come in two types: central-mixing and dry-batching.

A central-mix plant combines all the components (except certain admixtures that are put in at the last minute) in a stationary mixer. After mixing is complete, the concrete is discharged into trucks for delivery to the job site.

A dry-batch plant puts the dry ingredients into concrete trucks. Water is then added and the mixing takes place in the truck drum. There is a prejudice in some quarters against dry batching, but with good equipment the method produces concrete every bit as good as central mixing. Dry batching comes into its own on long hauls; the driver can wait till the right moment to add water, making sure the concrete arrives fresh at the job site. That advantage turns into a minor drawback if the job site lies near the plant. Then the drive may be too short for complete mixing, and it may be necessary to let the truck mix for a few minutes after arrival on site.

For proper mixing of dry-batched concrete, the truck's mixing drum has to turn at high speed – faster than the so-called agitating speed used after mixing is complete. ASTM C 94 recommends 70–100 revolutions at mixing speed. In the UK, Blackledge (1980, p. 3) recommends at least

100 revolutions. You can mix too much, however. The concrete should be discharged before the drum has turned 300 times.

ACI C 94, which uses the term truck-mixing instead of dry-batching, recognizes a third category of ready-mixed concrete called shrink-mixed. Shrink-mixed concrete is partially blended in a stationary mixer and then put into a truck were the mixing is completed.

Whatever the mixing method, no concrete maker has perfect control over the process. Even if the dry ingredients are batched with great accuracy, variations in the moisture content of the aggregates cause workability to vary from batch to batch. Wash water left in the mixing drum has a similar effect. The resulting changes in workability can cause problems in concrete finishing, with wetter batches setting up later than drier ones. Some floors, especially those with tight flatness tolerances, require closer control over water content and workability than the typical ready-mix plant can achieve.

One solution to that problem is to order the concrete slightly stiffer than desired, and then adjust it on site by adding water. Some specifications forbid the addition of water on site, but such a prohibition is hard to justify. Site-added water is safe as long as the following rules are followed:

- The addition of water must be under the control of one responsible person.
- The batch ticket must show the amount of water that can be added without jeopardizing strength and other specified properties.
- The amount of added water must be recorded.
- The concrete must be well mixed after the water is added.

A common mistake is to add water with mixing it improperly. After water is added, the drum should make at least 30 revolutions at mixing speed.

Mixing on site

Site-mixed concrete used to be common, but ready-mixed concrete has largely replaced it throughout the industrialized world. Nowadays, even the builders of small house and garage slabs are likely to order their concrete from a ready-mix plant.

Still, site mixing hangs on at both ends of the project-size scale. It makes sense for some very small jobs where the quantity of concrete needed is less than a truckload. Site mixing can also work well on jobs where concrete is used so slowly that it would take an hour or more to pour out a standard concrete truck. Some thin toppings and repair projects fall into that category.

Figure 10.1 Temporary batch plant near the job site

At the other extreme, some contractors set up mobile batch plants on huge jobs. These plants are effectively the same as ready-mix plants, but are made to be dismantled and moved from job to job. They are often used in remote areas where the local concrete suppliers, if any, cannot supply the quantities (or quality) desired. Figure 10.1 shows a mobile batch plant set up in a rural part of New York state for a distribution centre of 50 000 m² (500 000 ft²).

On small jobs, site-mixing is often associated with rough batching and a casual attitude toward quality, but it does not have to be. Some builders set up scales and weigh the components with as much care as any ready-mix producer. Others batch by volume. Though some authorities deprecate the practice, careful volume-batching can produce perfectly satisfactory floors. Batching by volume is the norm for sand–cement screeds.

Ready-mix plants and mobile batch plants use bulk cement, but smaller site-mixing operations generally get cement in bags. Each batch should contain a whole number of bags. In the UK a standard bag holds 50 kg. The American standard is 94 lb, which seems a strange amount till you measure its volume – 1 ft³.

Almost any mixer in good condition will serve for the highly workable concrete normally used in floors. Cement–sand screed mixes are much stiffer, however, and require a forced-action mixer.

11

Transporting and Placing Concrete

After concrete has been mixed, it must be transported and placed. Transportation is the moving of concrete from the mixer to the point of pour. Placement is putting of concrete in its final location, ready for the next step, finishing (see Chapter 17).

In some ways, placing concrete is easier in floors than in columns or walls. Floors are usually thin, so we need not worry much about horizontal cold joints. Reinforcement is usually simple, and some floors lack it completely. Driven by the need for finishability, concrete for floors is usually highly workable.

In other ways, floor construction imposes unusual demands on concrete placement. In this chapter we shall look at:

- transporting concrete;
- slab layouts;
- side forms;
- tools for placing concrete;
- compaction.

Transporting concrete

As a general rule, we should transport concrete as little as possible. We would not want to haul concrete across town in a ready-mix truck, transfer it to a dumper and then deposit it in a pump. Too much handling causes segregation – the partial unmixing of the components. Nevertheless, concrete always has to be transported, even if only from one part of the job site to another. The following methods are common in floor construction:

- ready-mix trucks;
- tipper (dump) trucks;
- dumpers;
- wheelbarrows and handcarts;
- pumps;
- cranes.

Ready-mix trucks

These transport concrete from ready-mix plants, and are also used with on-site batch plants. The typical model holds between 5 and 10 m^3 (6 and 13 yd^3). It has a rotating drum with spiraling internal fins. When the drum rotates one way, the fins push the concrete down, mixing it. When the drum reverses its direction, the fins push the concrete up and out for discharge. The drum should never stop rotating for more than a few seconds as long as it contains concrete. Rotation can be fast or slow. Fast rotation is called mixing speed, and is used to mix the concrete and to blend in extra water and admixtures. Slow rotation is called agitating speed, and keeps the concrete mixed.

At dry-batch plants, the concrete is actually mixed in the truck's drum, requiring about 100 revolutions at mixing speed. At central-mix plants, the concrete is mixed before it enters the truck, eliminating the need for mixing speed unless something is added later. According to ASTM C 94, *Standard Specification for Ready-mixed Concrete*, the number of revolutions should not exceed 300, whether at mixing or agitating speed. If extra water is added on site, the drum should spin at least 30 times at mixing speed.

Wherever possible, ready-mix trucks should discharge their loads directly into the forms (see Figure 11.1). This saves so much time and money it often justifies major changes such as leaving out walls and building ramps.

Ready-mix trucks are heavy and easily tear up subgrades and sub-bases, especially where they have to make tight turns. On some sites, it may be necessary to lay gravel or crushed rock roads for the trucks.

On big pours, you need enough trucks, properly spaced, to maintain a steady supply of concrete. It is no good having three or four trucks queue up because they are arriving faster than the concrete can be placed. On the other hand, it is equally bad to run out of concrete and have to wait

Figure 11.1 Discharging concrete straight from the ready-mix truck into the forms

as the trucks go back for reloading. The timing of concrete delivery is more critical in floor construction than in most other concrete work.

Most suppliers will, for reasons of economy, send their trucks loaded to capacity, whether that capacity is imposed by the truck's size or by the weight limits for public roads. (In some places the suppliers do not seem to worry much about weight limits.) Generally you want full loads, except for day's last "make-up" load. But in a few cases smaller loads are preferable, even if you have to pay extra for them. When laying a thin topping or making certain repairs, you may use concrete so slowly that a full truckload would take two hours or more to pour out. Ordering partial loads, well spaced, helps ensure the concrete is still fresh when it comes out of the truck.

If a job is too small or too slow to justify a ready-mix truck, ready-mix concrete may still be available. Some suppliers offer small ready-mix trailers that can be towed behind a car or van.

Tipper trucks

Tipper trucks are called dump trucks in America, these can deliver large quantities of concrete and discharge it fast. But they are severely limited, compared to ready-mix trucks, because they do not agitate, much less mix, the concrete. Their use is confined to very short hauls of site-batched concrete, and slightly longer hauls of very stiff mixes. Using them under any other conditions invites segregation.

Dumpers

These are called buggies or power buggies in America, these miniature tipper trucks transport concrete for short distances (Figure 11.2). The typical dumper holds 0.3 to 0.8 m³ (0.4 to 1 yd³). Some discharge their load by gravity. Others rely on hydraulics to tip the load and offer a more controlled discharge.

Dumpers are often a good choice for moving site-mixed concrete from the mixer to its final location. They are also used to transport ready-mixed concrete from the truck to the slab on sites where the bigger trucks cannot get in.

Because they do not agitate the concrete, dumpers should not be used for long hauls.

Figure 11.2 Using a power dumper (buggy) to place concrete for a thin, bonded topping

Wheelbarrows and handcarts

These are common in countries where labour is cheap. Even in high-wage areas, wheelbarrow and handcarts find use on small jobs and in congested areas.

The typical wheelbarrow has an effective capacity of only $0.03\,m^3$ ($1\,ft^3$), which severely limits its efficiency. In contrast, a handcart holds about $0.2\,m^3$ ($7\,ft^3$). That is much more than a wheelbarrow can carry, but it makes the handcart correspondingly harder to push. Handcarts are all but useless on sloped ground.

All hand-pushed vehicles repay forethought with regard to access, travel distance and smoothness of the path on which they run. Efficiency falls if workers have to negotiate potholes, muddy patches and narrow doorways.

Pumps

None of the methods discussed so far will deliver concrete to an elevated slab. Upper-storey pours usually require a pump or crane. Pumps (though not cranes) are also widely used for floors at ground level.

A pump delivers concrete through either a pipeline or a placing boom. A pipeline consists of short lengths of metal pipe that are assembled on site before the pour begins. Each section is about $3\,m$ ($10\,ft$) long. Common diameters are 100 and $125\,mm$ (4 and $5\,in$). The 125-mm (5-in) size may be needed where the concrete contains 40-mm ($1\frac{1}{2}$-in) aggregate. The last section (the "elephant's trunk") is made of rubber so it can be moved from side to side. The pour starts at the point farthest from the pump. As the pour proceeds, workers remove sections of pipe. While a pipeline can be used on almost any site, it restricts the work somewhat. Wide-bay pours are difficult, and work has to stop every few minutes as the line is shortened.

The alternative to a pipeline is a placing boom (see Figure 11.3), consisting of permanently assembled pipes that are moved by hydraulic power. Booms are easier to use than pipelines and allow greater flexibility in pour size and shape. Many sites lack the room for them, however. They work best on outdoor pours.

Builders like to brag about how far and how high they pump concrete. Nevertheless, it is always a good idea to keep the distance as short as possible.

Figure 11.3 A placing boom works well on outdoor pours

Most pumps need to be primed with cement-rich grout. Some builders use cement and water mixed right at the pump mouth. Others order about $1\,m^3$ ($1\,yd^3$) of grout from the ready-mix supplier. Either way, the grout should be thrown away after it has passed through the pump and pipeline. It should never be incorporated into the floor slab, because it has properties much different to those of the specified concrete.

Pumping sometimes requires a slightly different concrete mix than would otherwise be used. A pump mix often contains a bit more fine aggregate (3–5% more), and sometime more cement. The properties that make concrete easy to pump also make it easy to finish, so a mix designed for pumpability usually works well in a floor.

Pumping causes concrete to lose workability. The loss varies and is less than some people suppose. For most mixes in most pumps, the loss of slump does not exceed 20 mm (1 in).

Because pumping affects workability, arguments often arise over where to take slump tests. Some floorlayers request, and some specifications demand, that tests be made where the concrete leaves the pipeline. That is risky, however, because a bad batch may go undetected till a substantial quantity of it is already in the pump. It is better to specify a slightly higher slump and test it where the concrete enters the pump.

Cranes

These are used to transport concrete on high-rise buildings. They are almost never used on ground-level pours. The crane lifts a dump-bucket that typically holds about 3 m^3 (4 yd^3). Provided the space above the floor is clear, cranes can put concrete exactly where it is needed and work well no matter what the size or shape of the pour.

Slab layout

A small floor can be placed as a single pour, but bigger floors are normally divided into slabs. The slabs are cast on different days and are separated by construction joints.

Some designers control the slab layout, while others leave it to the floorlayer. Either way, it is an important part of the job because it affects all these properties:

- cost;
- schedule;
- ease of construction;
- crack control;
- curling;
- surface regularity.

There are four basic layouts:

- chequerboard;
- narrow strips;
- wide strips;
- large bays.

Chequerboard construction

This is almost obsolete, but a few specifications still call for it. The floor is divided into small square (or almost square) panels. Every other slab is cast first (think of the dark squares on a chequerboard), and the infill slabs are cast later (Figure 11.4).

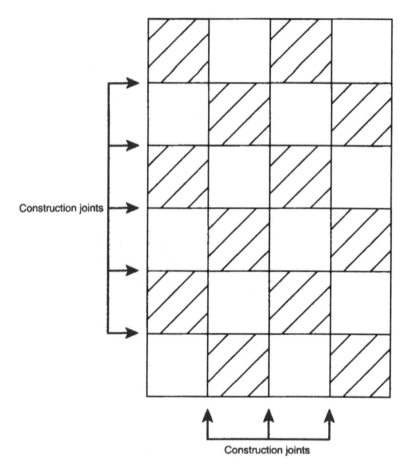

Figure 11.4 Chequerboard construction

Chequerboard construction had some merit back in the days when floors were laid by hand and concrete saws were rare. There is little to recommend it today, however. Modern practice leans toward longer and sometimes wider slabs, fewer construction joints and sawcut joints to control cracks.

Narrow-strip construction

This divides the floor into long narrow strips, each up to 6 m (20 ft) wide. Most builders lay every other strip first, and the intermediate strips, called infills, later. But there is nothing wrong with laying the strips consecutively if that proves more convenient.

Thanks to its narrow width, each strip is well protected from longitudinal cracks. Transverse cracks can be troublesome, however. Most designers employ sawcut joints, reinforcement or post-tensioning to control transverse cracks.

It is usual to cast an edge strip 0.6–1 m (2–3 ft) wide next to each side wall. No edge strips are needed at the slab ends unless the floor is post-tensioned. In that case, a temporary gap at one or both ends leaves the anchors accessible for stressing. If the floor is cast before the walls, no edge strips are needed.

Narrow-strip floors can be struck off with a wide range of tools: hand straightedges, twin-beam vibrators and truss vibrators.

Narrow-strip construction was the norm in the UK in the 1980s and early 1990s. The industry there standardized on a slab width of 4.5 m (15 ft) and made vibrating screeds and wire mesh to suit that dimension.

That has changed. The growing use of laser screeds and large-bay construction has relegated narrow-strip construction to a shrinking subset of the UK market.

There is, however, one area where this method still reigns, all over the world. Superflat floors are almost always laid in narrow strips (Figures 11.5 and 11.6).

Figure 11.5 Narrow-strip construction is standard for superflat floors

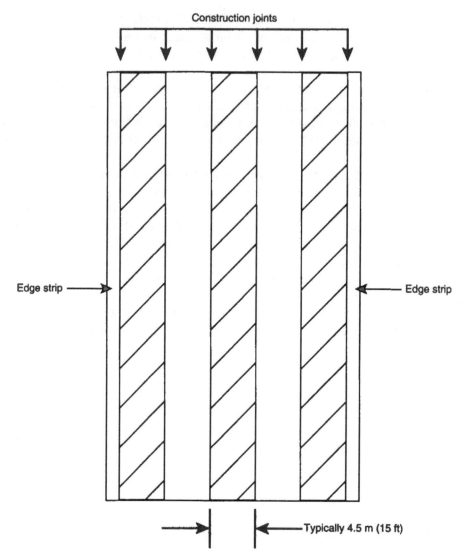

Figure 11.6 Narrow-strip construction

Wide-strip construction

This method divides the floor into long strips over 6 m (20 ft) wide. Strips vary in width, but each typically takes one whole column bay and does not exceed 15 m (50 ft) wide (see Figure 11.7).

Compared to narrow-strip construction, the wide-strip method costs less and goes faster, mainly because it needs less formwork. On the downside,

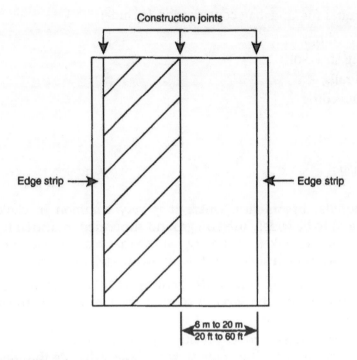

Figure 11.7 Wide-strip construction

it provides less control over surface regularity. While narrow strips can be laid to almost any tolerances (not that all are), wide strips should not be specified much higher than $F_F 50/F_L 30$. But that is good enough for most purposes.

Too wide for hand strike-off, these slabs are usually placed with a truss vibrator. Nowadays some floorlayers use laser screeds to place wide strips. But the laser screed really comes into its own on even wider placements, discussed next.

Large-bay construction

This term encompasses several different techniques for laying big floors with few construction joints. Some floorlayers routinely place over 3000 m³ (30 000 ft³) in a days work. Single pours have exceeded 10 000 m³ (100 000 ft³). The length and width of the pour is limited only by the availability of concrete and the time needed for placing and finishing.

Techniques for large-bay construction include:
- wet screeding;
- floating strike-off;
- screed rails;
- laser screeding.

Wet screeding

This is popular in America, where it is very common in elevated-slab work. It used to be widely used on ground-supported industrial floors, but laser screeding has taken over a big part of that market. Each floorlaying crew develops its own style, but the following steps are typical:

1. Set grade stakes on 5-m (15-ft) centres in two directions. These are stakes driven into the ground so that their tops lie at the specified finish-floor elevation.
2. As soon as concrete arrives on site, lay narrow strips of concrete along the lines connecting the grade stakes. These strips are the wet screeds. They generally run in just one direction.
3. Use a straightedge (sometimes timber, but preferably extruded metal tubing) to level the wet screeds with the grade stakes.
4. Before the wet screeds harden, pour concrete between them and strike-off, using the wet screeds as side forms.
5. Remove the grade stakes (see Figure 11.8).

Wet screeding is cheap and fast – attractive properties – but is has at least two drawbacks. First, it requires much skill. An inexperienced gang can get into deep trouble. Second, it offers little control over surface regularity. Exceptions exist, but the typical contractor laying to wet screeds is hard pressed to beat $F_F 30/F_L 20$ on ground-supported floors, or $F_F 25/F_L 15$ on elevated slabs.

Floating strike-off

This resembles wet screeding, but there are no grade stakes or wet screeds. A worker continually checks the elevation of the strike-off with an optical level or a laser. Some specialists do a good job with this method, but it is not for beginners. As practised in the UK, floating strike-off normally relied

Transporting and Placing Concrete

Figure 11.8 Wet screeding

on highly workable concrete made with superplasticizer. That method has largely given way to laser screeding, however.

Advantages and drawbacks are the same as for wet screeding.

Screed rails

These allow large bays to be placed like narrow strips. The screed rails – beams or tubes set at finish-floor elevation – support the striking-off tools as the side forms do in narrow-strip construction.

Most screed rails are temporary. Timber strips or steel pipes are chaired up above the ground (for ground-level floors) or soffit form (for elevated floors). Once they have served their purpose and before the concrete has fully set, the screed rails are pulled up and the void filled with fresh concrete.

Some screed rails are permanent. One type is the metal screed key, a thin steel stamping that creates a keyed joint. Though its main purpose is to make the joint, it can also serve as a screed rail for lightweight strike-off tools. Metzger (1983) warns against screed keys where the floor supports industrial traffic, and my experience confirms that.

Figure 11.9 Using a laser to set precast concrete screed rails

Another permanent type is the precast concrete rail, often identified by the trade name Permaban. Concrete rails are set in pads of concrete and can serve as edge forms as well as screed rails (Figure 11.9). They can support industrial traffic if provided with dowels or tie bars for load transfer.

Laser screeding

Since its introduction in the late 1980s, this has become the standard method for large-bay placement in many parts of the world. The technique depends on a device called a laser screed, a term almost synonymous with

Figure 11.10 Laser screed

the patented and trademarked Somero Laser Screed. A laser screed – or, as we might as well say, the Laser Screed – is a heavy machine that strikes off concrete with a horizontal auger. As the auger passes over the fresh concrete, a system of sensors and motors maintains a fixed distance below the beam of a laser level. Each pass levels a patch of about $20\,m^2$ ($200\,ft^2$). Then the self-propelled machine rolls on to the next patch(Figure 11.10).

Laser screeding gives excellent control over floor levelness without the need for narrow slabs and accurate side forms. It has enabled floorlayers to make huge pours while maintaining a degree of surface regularity that used to demand narrow-strip construction. Flatness and levelness numbers of $F_F 50/F_L 35$ are routinely achieved, and some floorlayers regularly meet $F_F 60/F_L 40$ on pours that exceed $3000\,m^2$ ($30\,000\,ft^2$).

As useful and popular as laser screeding has proven to be, it has a few drawbacks. It requires costly equipment and a trained operator. The standard machine weighs too much to use on most elevated floors, and the lighter models that do work almost anywhere cannot match the standard machine in productivity. Probably the most far-reaching limitation is the difficulty of laser screeding floors that contain reinforcement or post-tensioning tendons. There are ways to do it – light reinforcement can be installed at the last minute, or bridges can be set up to let the laser screed ride above the steel – but most laser-screeding contractors prefer floor designs that do not require them to work around embedded steel. This preference goes a long way toward explaining the growing popularity of unreinforced and steel-fibre reinforced floors.

Side forms

Some floors are cast right up to the walls on all sides. All other cast-in-place floors need side forms, which must meet two requirements.

First, they must be strong enough to support the lateral pressure of the fresh concrete and the weight of the tools used for striking-off. Strike-off tools vary widely, in wide-strip construction, the main tool is likely to be a truss-type vibrating screed with a mass of several hundred kilogrammes (several hundred pounds). Narrow-strip construction relies on lighter tools – either short vibrating screeds or hand straightedges. Laser screeding is easy on forms because the screed machine, though very heavy, never rides on the side form. It stops short of the form, leaving the edge to be worked with lightweight hand tools.

Second, the side forms must be accurate in elevation because they help determine a floor's surface regularity. Even in laser screeding, where the screed machine does not rely on the forms to level the floor, form elevations determine surface regularity at the construction joints. The need for accuracy varies according to the flatness and levelness requirements for the finish floor.

Deacon (1986, p. 12) recommends a tolerance of ±2 mm (±0.08 in) from datum for general work, and ±1.0 mm (±0.04 in) or less for superflat floors. But those are seldom achieved in actual site practice. They probably represent the best work rather than the typical.

Most builders set forms to grade using an optical level or laser. It is hard to do much better than ±3 mm (±1/8 in) with such instruments, but that is good enough for most floors.

Forms for superflat floors are often set with the same tools, but typically get a second check from a different tool – either a long straightedge or a self-contained electronic level (see Figure 11.11).

Side forms are made of steel, wood or concrete.

Steel forms (see Figure 11.12) are rugged but not always straight. One drawback to them is their short length – some are only 2 m (6 ft) long.

Wood forms are very common in America. They offer great flexibility, since they can be sawn to length and width, and are easily drilled for dowels or tie bars at any spacing. Some builders use low-quality timber for forms, but that need not be so. Figure 11.11 shows a form made of straight, clear timber, ready for use in a superflat floor. Figure 11.13 shows a form made of rougher stock, heavily braced because the slab is thick.

Transporting and Placing Concrete   **191**

Figure 11.11 Using an electronic level to check a timber form on a superflat job

Concrete forms (see Figure 11.9) are precast rails made of high-strength concrete. They are laid on pads (or sometimes a continuous strip) of concrete, and are left in the finished floor.

Tools for placing concrete

Hand labour still moves concrete into its final position in most floors. The best tools for this job are square-ended shovels and come-alongs. A come-along is similar to a garden hoe, but with a wider blade.

192 Design and Construction of Concrete Floors

Figure 11.12 Steel form

Figure 11.13 Heavily braced timber form

Some workers place concrete with rakes, but this practice can cause segregation.

Compaction

After concrete has been placed in the forms it is still not quite ready to be finished. It needs to be compacted first. Compaction eliminates air pockets from the fresh concrete.

When concrete is poured it contains large amounts of entrapped air. This differs from the entrained air that is deliberately introduced to help concrete resist frost. Entrained air consists of millions of microscopic bubbles, each between 0.01 and 1 mm (0.004 and 0.04 in) across. Entrapped air voids are less regular and typically more than 1 mm (0.04 mm) across.

The amount of entrapped air varies with the concrete's workability. Blackledge (1980B, p. 5) says concrete placed at a 75-mm (3-in) slump contains about 5% air by volume. If you reduce the slump to 25 mm (1 in), the air content rises to about 20%. High-slump flowing concrete, made with superplasticizer, entraps little air and is generally not compacted. It could be considered self-compacting.

We need to reduce the entrapped air to about 1% of the concrete's volume, for these four reasons:

- Entrapped air weakens concrete. The rule of thumb is that each 1% of entrapped air causes a 5% reduction in concrete strength.
- Entrapped air causes fresh concrete to settle by a large and highly variable amount. This leads to plastic-settlement cracks and poor surface regularity.
- Entrapped air makes the hardened concrete more porous, reducing its resistance to chemical attack.
- Entrapped air reduces the bond between concrete and reinforcing steel.

The most effective way to compact concrete and reduce entrapped air is vibration. Hand-tamping is still used on some floors, but vibration works better. Two kinds of vibrators are used on floors: pokers and vibrating screeds.

The poker vibrator, also called an immersion or internal vibrator, gets its name from the fact that the business end of the tool is stuck into the fresh concrete. It consists of a vibrating cylinder 25–75 mm (1–3 in) in diameter and 300–600 mm (12–24 in) long. Floorlayers use mainly the smaller versions, but big ones can be effective in thick slabs. Some guidebooks recommend that the poker be kept vertical when in use, but that rule is hard to follow in floor work. The right way to use a poker vibrator is lower it quickly into the concrete at points spaced 450–750 mm (18–30 in) apart. Leave the poker at each spot till the concrete around it is fully compacted, then pull it out slowly and move on to the next spot. You can tell the concrete is fully compacted because mortar shows at the floor surface and bubbles stop appearing. Try to avoid over-vibration because it can cause segregation, but if in doubt always err on the side of more vibration, because under-vibration causes more harm than its opposite.

Unlike pokers, vibrating screeds ride over the concrete and compact it from the top down. They are strike-off tools first and vibrators second, and so are described in Chapter 17. Vibrating screeds can compact concrete to a considerable depth – up to 300 mm (12 in), according to some. However, it is not easy to gauge the depth of vibration. The surface always looks well vibrated because the action is greatest there. But if the screed moves forward too fast, lower depths may go uncompacted. The only solution is to advance the screed slowly.

Laser screeds, though often put in a separate category from vibrating screeds, work the same way when it comes to compaction.

All the concrete in a floor should be compacted, but the edges call for special attention. The concrete there needs all the strength and density it can get. And if the slab edges hold post-tensioning anchors, compaction is essential to prevent the anchors breaking free when the tendons are stretched.

Specifying compaction

Designers usually specify compaction by description. A typical specification requires that all concrete be thoroughly compacted by vibration, and goes on to list the acceptable kinds of vibrators. Routine compliance testing almost never occurs, though suspect concrete can be cored and examined for voids.

A performance specification for compaction is within reach, however, even if rarely used. It comes from highway construction, where poor compaction is a serious risk because of the stiff concrete mixes used. *Concrete Construction* (1989) describes a method for measuring compaction with a nuclear density gauge.

The next step

After concrete has been transported, placed and compacted, the next step is finishing. Because finishing is closely tied to the properties of the concrete surface, it is discussed in Part V. In the next chapter we shall jump forward to curing, which takes place after the concrete has been finished.

12

Curing

Curing is the process of keeping concrete damp so that hydration, the reaction of cement with water, can continue. Though concrete sets in a few hours, it continues to gain strength for a long time if kept moist. The strength gain is rapid at first but slows over time. If concrete dries out too soon, it will achieve only a fraction of its potential strength and durability.

Curing is simple and often neglected – but of huge importance. All concrete structures benefit from good curing, but floors are especially demanding because they have a large area exposed to drying, and also because that exposed area becomes, in many cases, the wearing surface. A poorly cured floor may experience any of the following defects:

- crazing;
- dusting;
- low resistance to wear;
- premature spalling at joints;
- low resistance to chemical attack;
- low resistance to frost damage.

Figure 12.1 illustrates what a difference curing can make. It shows the results of Chaplin wear tests on two samples that differ in only one way: the one on the left was left to dry out, while the one on the right was cured under polyethylene sheet.

Despite curing's great importance, it sometimes gets more credit than it deserves. Some people believe curing reduces the problems, including cracking and curling, caused by long-term drying shrinkage, but the evidence does not support that belief. Unless a slab stays moist forever – an impossibility in most buildings – curing only delays and does not reduce drying shrinkage.

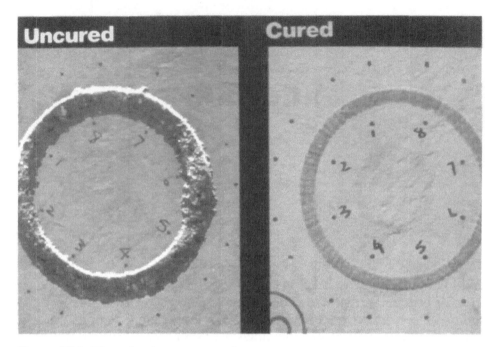

Figure 12.1 Effect of curing on wear resistance

Curing methods

The many methods (Figure 12.2) fall into four broad categories:
- flooding;
- absorbent materials kept moist;
- waterproof sheet materials;
- curing compounds.

Wet curing

Some specifications call for wet curing, but this term has been applied to so many different methods as to rob it of all effective meaning.

To some people, wet curing encompasses almost every method other than the use of a curing compound. Others reserve the term for methods that involve the addition of water during the curing period – a definition that rules out waterproof sheet materials and curing compounds. A few strict constructionists equate wet curing with flooding. But in one sense all curing is wet curing, since the whole point is to keep the concrete moist.

Curing 197

Figure 12.2 Curing methods: acrylic resin on left, polyethylene sheeting on right, no curing (air drying) at bottom

Those whose definitions call for added water often regard wet curing as superior to other means – such as waterproof sheet materials and curing compounds – that introduce no water but merely hold in the water the concrete started out with. The theory is that added water replaces the water consumed in the hydration process, eliminating the risk of a water deficit. That sounds reasonable on first hearing, but in truth the concrete used in floors always contains far more water than the minimum needed to hydrate the cement. There is always enough water in the concrete mix, provided we keep it there.

Flooding

This cures concrete by keeping the surface covered with water. One way is by ponding, which involves building a short dam on all sides of the floor. Another way is by spraying, but that is rarely practical except on the smallest slabs. For good results, the spray must cover the whole floor and be kept on continuously for the duration of the cure.

Flooding can do a superb job, but is rarely seen these days in the UK and America.

Absorbent materials kept moist

This method relies on materials that absorb large quantities of water. The absorbent material is spread over the new floor and soaked with water. Under all but the dampest conditions, the materials must be wet again from time to time.

Loose absorbent materials such as sand, sawdust or straw are seldom used these days in the industrialized world, but still appear elsewhere. They tend to dry out fast and need to be sprayed with water often.

Absorbent sheet materials include hessian (Americans call it burlap) and building paper. Specialist floorlayers traditionally used them for high-end industrial floors, but the practice had declined by the end of the twentieth century. They do a good job if kept moist, and are almost useless if allowed to dry out. For that reason, it's risky to specify an absorbent sheet material without providing the close supervision and labour to keep it wet.

Absorbent sheet materials are sometimes combined with waterproof sheet materials. In that case, the absorbent material goes on the bottom and the waterproof sheet goes on top keep it from drying out. Some manufacturers make a combined product that consists of absorbent fabric bonded to a plastic sheet. Burlene is the best-known brand. Some builders make their own by covering the floor first with hessian, and then covering that with polyethylene.

Waterproof sheet materials

These work by reducing moisture loss from the floor surface.

By far the most common sheet material for curing is polyethylene film. It comes in clear, white or black. Any colour will do indoors, but white is best for slabs cast outdoors in hot weather because it keeps the concrete cooler. ASTM C 171 recommends nominal thickness of at least 0.10 mm (0.004 in or 4 mils), with an absolute minimum of 0.075 mm (0.003 in). Some floorlayers prefer thinner film – nominal 0.05 mm (0.002 in) – on the grounds that it conforms more closely to the concrete surface. But thinner film tears more readily.

The chief complaint about polyethylene is that it leaves light and dark streaks on the floor surface. Though they fade over time, the streaks start out obvious enough to make polyethylene a poor choice for floors that must present the best possible appearance. Coloured concrete – never used except where looks count – should never be cured under polyethylene.

If streaks are unacceptable but you still want to use sheet material, waterproof paper is another option. In America, the standard curing paper consists of two layers of reinforced kraft paper bonded stuck together with bituminous resin and reinforced with filaments.

To do their job, sheet materials must cover the whole slab and must stay in place throughout the curing period. Some specifications call for taping the seams, but hardly anyone does that. The usual practice is to lap the seams. Unless the floor is fully sheltered from wind, the laps should be weighted down with timber or reinforcing bars. If the cover is peeled back for any reason, it should be replaced as soon as possible.

Some floorlayers drench the floor with water before applying a waterproof sheet material. That can't hurt, but is probably not strictly necessary provided the curing starts as soon as possible.

Curing compounds

These are liquids that dry or harden to form a water-retaining membrane on the floor surface. They are usually sprayed on, but some can be applied by roller, brush or mop.

Economy is the main reason to use curing compounds. They usually cost less than other methods. But they have some practical advantages as well. You can apply them once and then forget about them. In contrast, all other methods require attention throughout the curing period to make sure the concrete stays moist. The ability to spray something on and walk away makes curing compounds attractive where schedules are short or supervision scarce.

Curing compounds have at least three drawbacks. First, the cheaper, lighter products just do not cure as well as other methods. You cannot expect an 18%-solids acrylic resin to retain water nearly as well as a sheet of polyethylene or water-soaked hessian. Second, curing compounds can interfere with some coatings and floorcoverings. To reduce that risk, some compounds, called dissipating resins, are designed to degrade and disappear over time. Third – and this applies only to industrial floors – some compounds show tyre marks more readily than does bare concrete.

The chemical industry makes a wide variety of curing compounds. Most of those used on floors fall into one or more of these categories:

- polymer resins;
- waxes;
- dual-purpose bondbreaker-curing compounds;
- liquid hardeners.

Polymer resins

The most widely used curing compounds consist of polymer resins in solvent. The traditional solvent was one or more aromatic hydrocarbons, but nowadays many products rely on water instead. Water-based compounds create less odor and are required by law in some places to keep down air pollution. Some of the resins used in higher-grade curing compounds are:

- styrene acrylate (often called acrylic);
- styrene-butadiene copolymer;
- chlorinated rubber (rare in the UK but an old favourite in America).

The resin contents vary from about 15 to 40%. Compounds at the higher end retain moisture better but cost more. Compounds at the lower end may not suffice where the concrete forms the wearing surface. Industrial-floor specifications often require that the curing compound contain "at least 30% solids". Compounds with lots of solids are often marketed as curing-sealing compounds, since they provide a membrane that lasts far beyond the usual curing period.

Some resin compounds include a fugitive dye to confirm the coverage. The dye fades over time. Other compounds contain white pigment to keep the floor cool in hot sunshine. Unlike fugitive dye, the white can stick around a long time and should not be used where people might object to its look.

Waxes

Waxes are the curing compounds of choice for coloured concrete. The wax is dyed to match the concrete. A wax curing compound on its own cannot reliably colour the floor, but works well in conjunction with pigmented concrete or a coloured dry shake.

Colourless waxes can be used on ordinary concrete floors.

Dual-purpose bondbreaker-curing compounds

These are unique to tilt-wall (also called tilt-up) buildings, in which the concrete floor forms the casting bed for the concrete wall panels.

Tilt-wall construction creates several challenges with regard to curing. Slabs are cast outdoors, exposed to sun and drying winds. That creates the need for a highly effective curing method. But tilt-wall slabs also need to be coated with a bondbreaker so the wall panels will not stick. That has led to the development of dual-purpose products meant both to prevent bonding and to cure the floor slab. Those products fall short of the ideal, however. They all work fine as bondbreakers, because no contractor can tolerate a failure in that area. They do not all work so well as curing compounds. Some are recommended only for mild conditions. And even those recommended for more severe conditions may not match the water-retaining abilities of the highest-grade, pure curing compounds.

For tilt-wall floors that get coverings, the dual-purpose compounds are undoubtedly good enough. But for exposed industrial floors, especially those laid in dry or windy conditions, consider curing the floor by some other method, such as polyethylene or wet hessian, and then applying a pure bondbreaker after the curing period. Not everyone will like that advice, because it delays the job a few days and a short schedule is one of the main selling points for tilt-wall construction. But to do less is to run the risk of an inadequately-cured floor surface.

Liquid hardeners

These are not true curing compounds because they do not form a membrane that retains water. However, they deserve mention here since they are sometimes marketed as curing compounds or as alternatives to curing.

Liquid hardeners consists of salt – usually silicates but sometimes including siliconates – dissolved in water. Sodium silicate and magnesium fluorosilicate are perhaps the most common. The salts react with unhydrated lime near the concrete surface. As their name implies, liquid hardeners can, under certain conditions, make concrete harder and more durable. To that extent, they may compensate for inadequate curing. But if curing is defined as the process of keeping concrete moist, so hydration of the cement can continue, hardeners do not cure.

Specification of curing compounds

Curing compounds would seem to lend themselves to specification by performance.

In the UK, they are typically specified and classified by the efficiency index. In the standard test, four mortar specimens are cast in special moulds. Three specimens get the curing compound, while the fourth is left bare.

All four specimens are stored under controlled temperature and humidity. After 72 hours, the specimens are weighed to determine how much water they have lost. The index is calculated from this equation:

$$E = \frac{\left[100\left(W - W_1\right)\right]}{W}$$

where:

E = efficiency index, in %
W = percentage loss of water from bare specimen
W_1 = percentage loss of water from cured specimens

A test result of 100% would mean the cured specimens lost no water at all. A result of 50% would mean the cured specimens lost half as much water as bare concrete.

British chemical suppliers typically offer curing compounds in two grades: a run-of-the-mill compound with efficiency index of 75% and a better product, more suitable for floors, measuring at least 90%.

Americans use a similar test, standardized in ASTM C 156, but interpret the results differently. There is no bare specimen to serve as a control. The test result is the mass of the water lost in 72 hours, divided by the area of exposed surface. In this test, unlike the efficiency index, lower numbers mean better performances. ASTM C 309, *Standard Specification for Liquid Membrane-forming Compounds for Curing Concrete*, considers a curing compound acceptable if moisture loss does not exceed 0.55 kg/m² (0.11 lb/ft²) in 72 hours. Another standard, ASTM C 1315, imposes a tighter limit – 0.40 kg/m² (0.08 lb/ft²) – on special curing-and-sealing compounds.

That all sounds good, but there are serious pitfalls in using performance specifications for curing compounds. For one thing, some standard specifications are not very strict. This is particularly true of ASTM C 309, which seems devised more to include every manufacturer's products than to enable designers to specify the best-performing compounds. Another problem is that curing compounds are tested under standard conditions, which can vary greatly from what a floor actually experiences. Test method ASTM C 156 calls for applying the compound at 5.0 m²/l (200 ft²/gal), and curing at constant 23°C (73°F) temperature and 40–60% relative humidity. In real life, application rates are variable and hard to control, and the weather can range from dry desert heat to foggy cold. A curing compound that works in northern Europe, in March, might prove worthless in the Mojave Desert, in July. Standard specifications do not address that.

The solution is to keep on using the performance specifications but not to rely solely on them. Under severe drying conditions we need to specify the most effective curing compounds, either by brand name or by imposing performance limits tighter than those found in the national standards. And we need to supplement the standard specifications with descriptive instructions, whether spelt out in the contract documents or included in the manufacturer's instructions. Application rates must be tailored to the surface texture; float-finish or broom-finish floors need more compound – sometimes much more – than trowelled concrete. The main thing is to make sure the compound at least forms a complete film over the floor. If it looks speckled, the floor needs more curing compound.

How to choose a curing method

It helps to ask a few questions.

Is low cost important? If so, the choice will probably come down to polyethylene sheet versus a polymer-resin curing compound.

Will the floor get a coating or topping? If so, stay away from curing compounds unless you confirm they will not interfere with bonding.

Does appearance matter? If so, an absorbent sheet material kept moist may be the best choice. Polyethylene sheet leaves marks that some users object to. Curing compounds may create light and dark spots, and some of them show tyre marks.

Will the floor have to resist heavy industrial traffic? If so, this is not the job on which to compromise curing effectiveness. Choose flooding, sheet materials or only a top-grade curing compound.

Will the job get close supervision? If not, a curing compound may be the safest choice. It needs attention just once, at the time of application, and can be safely ignored from then on.

Will the floor be exposed to extreme drying? If so, use waterproof sheet materials or a top-grade curing compound. Methods that rely on adding water are risky here unless someone pays very close attention to make sure the floor does not dry out. Note that extreme drying conditions are not limited to hot, dry climates. They also occur inside artificially-heated buildings.

Timing

There is little dispute about when curing should begin. The sooner, the better. On floors that do not include sawcut joints, curing materials can and should go on right after the final trowel pass. Sawcut joints complicate the issue, because no one wants to remove curing materials for sawing. If the sawing can be done within an hour or two of final trowelling – as is often the case with modern early-entry saws – it is usual and reasonably safe to saw the floor first. But if the sawing must wait hours or overnight, then curing should start first even if they make more work later. On big pours in dry conditions, curing should start before the whole slab is finished.

No consensus exists on how long curing should last, but one fact is clear: you cannot possibly cure a slab too long. When in doubt, longer always beats shorter. Beyond that, recommendations range from a few days to a few weeks.

Murdock, Brook and Dewar (1991, pp. 212-213) suggest that a period as short as three days will suffice provided the concrete contains straight Portland cement (not pozzolans) and stays above 10°C (50°F). They may not have been thinking about floors when they wrote that, however.

Some authorities writing specifically about floors (American Concrete Institute, 1982, p. 53; Deacon, 1986, p. 15) recommend a 7-day cure. That is the closest thing we have to a standard duration.

There is evidence, however, that curing for longer periods offers benefits. Chaplin (1986) reports the results of one experiment in which stretching the cure from 7 to 21 days improved a slab's wear resistance by almost 10%. Longer curing is especially beneficial where concrete gains strength slowly. That applies where the concrete contains a high proportion of slag or fly ash, and where low-heat Portland cement is used. It can also apply to any concrete placed in cold weather – but in that case the evaporation rate is generally low, making a long cure less essential to success.

In short, a 7-day cure appears to be a reasonable choice under most conditions. But if you can stretch that to 14 or 21 days without wrecking the job schedule, do so. It cannot hurt, and may help.

Late curing

The right time to cure a concrete floor is at the start of its life, when the concrete is still saturated with its original mixing water. But if you miss that opportunity, late curing may be better than none. Chaplin (1986) describes an experiment in which curing was re-started after a 28-day interruption. The dry slabs were soaked with water and covered with polyethylene for 7 days, during which time their wear resistance rose substantially.

Part IV

Joints and Cracks

13

Cracks

Concrete floors often crack. A crack occurs when the tensile stress on the concrete exceeds the concrete's tensile strength. Many cracks are harmless, but some cause serious problems.

Designers and builders cannot guarantee crack-free floors. Some cracks can be prevented, however, and many cracks not prevented can be controlled so they do not make trouble for the floor user.

Because it costs money to prevent and control cracks, designers need to decide how far to go in dealing with each type of crack. Not every crack causes trouble, and the elimination of cracks is not always worth the cost.

This chapter discusses the following types of cracks, with suggestions on how to prevent or control each type:

- plastic-shrinkage cracks;
- plastic-settlement cracks;

  in plastic concrete

- crazing;
- drying-shrinkage cracks;
- thermal-contraction cracks;
- structural cracks.

  in hardened concrete

Plastic-shrinkage cracks

These occur when newly placed, still-plastic concrete undergoes severe drying. The cracks are widest at the top and seldom penetrate the full depth or extend the full width of the slab (Figure 13.1).

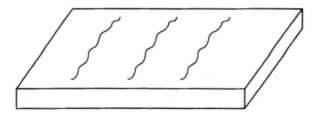

Figure 13.1 Plastic-shrinkage cracks usually stop short of the slab edge

Plastic-shrinkage cracking is most common in dry climates but can occur in generally humid areas like the UK, especially in high winds.

These cracks can look alarming but seldom do real harm. Some guides recommend beating the floor surface to fill the cracks in, but I have never seen anyone do that. The more usual remedy, not always effective, is to fill the cracks with mortar when the floor is floated.

Here are ways to reduce the risk of plastic-shrinkage cracks:

- Lay the floor indoors, where it is protected from wind and sunshine.
- When laying floors outdoors, use curtains or tents shield the concrete from wind.
- Avoid placing concrete on dry, windy days.
- Use misters upwind of the fresh concrete to raise humidity.
- Spray the fresh concrete with a monomolecular film to slow evaporation.

Plastic-settlement cracks

These occur when plastic concrete settles relative to fixed objects such as reinforcing bars. The crack appears directly over the fixed object. In bad cases, cracks appear over every bar, resulting in a near-perfect grid of cracks.

Plastic-settlement cracks are most common where reinforcing bars are thick and close to the surface. Thick slabs crack more than thin ones, because they settle more.

Here are some ways to reduce the risk of plastic-settlement cracks:

- Increase the cover – the distance from the floor surface to the reinforcing steel. Make sure this does not compromise the structural design.
- Use smaller reinforcing bars. By using more bars of smaller diameter, you can maintain the same amount of steel.
- Take steps to reduce plastic settlement, such as increasing the size of the coarse aggregate and eliminating gaps in the aggregate gradation.

Cracks

Crazing

This is a network of closely spaced, shallow cracks at the floor surface. Other names for it include map cracking (because it resembles the pattern of roads on a map) and alligator or crocodile cracking (because it vaguely resembles the fissures in the skins of those reptiles). In American practice, crazing describes a network with less than 50 mm (2 in) between cracks, while map cracks are separated by greater distances.

Crazing occurs when the floor surface shrinks relative to the concrete below it. This often happens very soon after the concrete has set. Contributing factors are a high evaporation rate and a smooth, burnished finish. Poor curing or a delay in curing is sometimes blamed, but crazing sometimes occurs even where the floorlayers take great pains to cure the concrete well.

Crazing is usually harmless. The narrower the craze cracks, the less likely they are to cause trouble (Figure 13.2). Narrower cracks are often associated with a closer crack spacing. This means a floor with a lot of craze cracks (that is, with a close crack spacing) may actually fare better than one with a few.

Wider craze cracks collect dirt and, in the worst cases, deteriorate under industrial traffic.

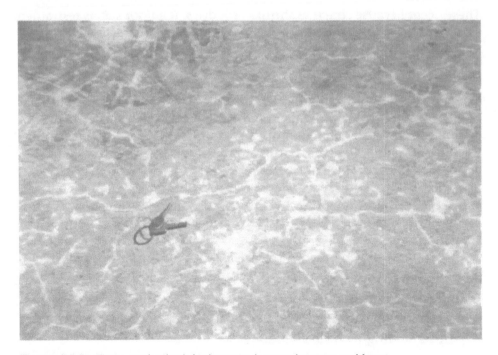

Figure 13.2 Craze cracks don't look pretty, but rarely cause real harm

Ironically, some of the best, most long-wearing industrial floors have craze cracks. The reason for that is those floors are power-trowelled to the smoothest possible finish, which greatly increases the likelihood of crazing.

Here are some ways to lower (but not eliminate) the risk:

- Do not float or trowel the floor surface while bleed water is present.
- Give the floor a broom or float finish (not a good choice, though, for industrial floors that must resist wear).
- Lay the floor indoors, where it is protected from wind and sunshine.
- When laying floors outdoors, use curtains or tents shield the concrete from wind.
- Avoid placing concrete on dry, windy days.
- Use misters upwind of the fresh concrete to raise humidity.
- Start curing as soon as possible.

Most crazing does not need repair. Where it does, options include grinding and the application of a resin seal.

Drying-shrinkage cracks

These occur because concrete shrinks as it dries. If a floor slab is restrained from moving freely, the shrinkage produces tensile stresses within the slab. If the tensile stress exceeds the concrete's tensile strength, which is not great, the slab cracks.

Unlike all the cracks described above, drying-shrinkage cracks usually extend over the full width and depth of the slab. They tend to start near points of restraint (at building columns, for example) and at planes of weakness in the slab (at cold joints, for example). They sometimes follow the lines of reinforcing bars, but by no means always.

It is hard to tell drying-shrinkage cracks from thermal-contraction cracks, described below. Both types are caused by restrained volume changes, and are identical in their effect on the floor. Some measures that control drying-shrinkage cracks work equally well against thermal-contraction cracks.

There are two general ways to deal with drying-shrinkage cracks: prevention and control. If prevention is the goal, the choices include:

- shrinkage-compensating concrete;
- pre-tensioning with steel tendons (mainly used in precast units);
- post-tensioning with steel tendons;
- steel fibres at very high dosages – 40-60 kg/m^3 (70–100 lb/yd^3).

Bear in mind that none of these preventive measures is foolproof.

Most floor designers do not even try to prevent drying-shrinkage cracks. They try instead to control them with reinforcement, joints or both. To control a crack is to manage its behavior in such a way that it causes little or no harm. Suspended floors, if designed according to one of the structural codes, normally contain enough reinforcement to keep cracks narrow and trouble-free. Ground-supported floors often lack the quantity of reinforcement needed to keep cracks tight and rely instead on joints to control where the cracks will occur. For more on this subject, see Chapter 16.

Though reinforcement and jointing are our primary crack-control tools, we have many other ways to reduce problems from drying-shrinkage cracks. Some methods work by reducing concrete shrinkage, while others reduce restraint on the concrete.

Here are some ways to reduce shrinkage:

- Use bigger aggregate.
- Avoid aggregates that show high drying shrinkage.
- Use less water in the concrete mix (but beware the effect on workability and finishability).
- Use less cement (but beware the effect on strength and finishability).

And here are some ways to reduce restraint:

- Isolate the floor from columns, walls and other fixed elements.
- Use a polyethylene slipsheet directly under the slab.
- Grade the sub-base with care.

Thermal-contraction cracks

These occur because concrete contracts as it cools. They look the same and have the same effect as drying-shrinkage cracks.

Thermal-contraction cracks can occur whenever the temperature falls, but are most common in the early days when the concrete is weak. Some evidence suggests that thermal contraction causes many of the early cracks often attributed to drying shrinkage. (Kass and Campbell-Allen, 1973)

Most of the methods that prevent and control (albeit imperfectly) drying-shrinkage cracks work equally well against thermal-contraction cracks. The exceptions are the mix-design changes that aim to reduce drying shrinkage. Adjusting the concrete mix will not substantially reduce thermal contraction.

We can prevent some thermal-contraction cracks by limiting temperature changes. This is hard to do after the floor is in use; no one heats or cools a building just to prevent floor cracks. We can, however, limit temperature swings during the floor's early life, when the concrete is weak and highly vulnerable to cracks. Here are some measures that have proven effective:

- Lay the floor indoors, after the building envelope is complete.
- Place concrete at night.
- Insulate the newly slab with blankets or straw (but not polyethylene sheets, which provide negligible insulation).

Structural cracks

These occur when a floor bends under load, creating tensile stresses that exceed the concrete's tensile strength. A structural crack may or may not denote a failure, depending on the floor design.

Structural cracks in suspended floors

These are normal. Being made of reinforced concrete, most suspended floors are designed to crack under load. Cracks are expected in areas of tension, and reinforcing steel is put there to resist that tension. A structural crack in a reinforced, suspended floor is nothing to worry about unless it grows wider than expected.

An unusually wide crack in a suspended floor may be the symptom of a very serious problem, however. The floor may be underdesigned or overloaded. Sometimes the cause lies with bad construction, such as missing or misplaced reinforcing steel.

How wide must a crack be before the alarm is raised? The answer varies with the design method. Some designers try to limit crack width to about 0.3 mm (0.01 in), but others allow much greater widths. American floors designed by the current strength-design method are allowed wider cracks than older floors designed by the working-stress method. In a pre- or post-tensioned floor, any crack wider than a hairline may be cause for concern. In any case, it is the floor design's job to predict crack widths and let other parties know what to expect (Figure 13.3).

Figure 13.3 Measuring crack width with a 50x microscope

Structural cracks in ground-supported floors

Unlike suspended floors, most ground-supported floors lack structural reinforcement and are designed not to crack under load. When a structural crack appears in a typical ground-supported floor, something has gone wrong.

Ground-supported slabs crack structurally for many reasons, including:

- loads that exceed design values;
- inadequate design;
- construction errors that result in a too-thin slab;
- subgrade failure.

Overloading during construction causes many cracks. Designers tend to focus on the user's loading, but some construction plant weighs far more than anything the user may have in mind. Tilt-wall construction requires a heavy crane to lift the wall panels, and it is often convenient to put that crane on the floor slab. Where this is allowed, structural cracks often appear around the crane footpads. Even light construction loads can cause cracks if applied before the concrete has gained much strength.

The way to prevent structural cracks during construction is to specify and enforce maximum loads, and to forbid any substantial loading till the concrete is strong enough to take it. The safest course is to allow nothing but foot traffic till the concrete has reached its design strength, but construction schedule don't always allow this luxury.

Though structural cracks in a ground-supported floor are evidence of failure, they are not always serious. An isolated crack seldom does real harm, especially if its cause was a temporary overload that is unlikely to recur. On the other hand, if cracks continue to form under normal loads, the floor will break up into every smaller pieces and eventually become unusable. Fortunately, this is rare.

Crack repair

The time to think about crack repair is before the floor cracks. To prevent (well, reduce) arguments, every floor specification should include a section on crack repair. It should state the conditions under which a crack needs repair, and it should spell out the allowable method or methods of repair.

Many cracks – perhaps most – do not harm. Cracks rarely need repair if hidden under floorcoverings or exposed only to foot traffic. Even in an exposed industrial floor, a crack generally need not be repaired unless it meets one or more of these conditions:

- Its width exceeds 1mm (1/32 in) and it is exposed to hard-wheeled traffic.
- It spalls under traffic, regardless of width.
- It shows differential movement under traffic.

Note that all three condition are related to traffic. Even in a busy warehouse or factory, cracks outside the main traffic areas can usually be left alone.

Some users have special requirements, however. In a food or pharmaceutical plant, and in some electronics plants, concern for cleanliness may force the sealing of every visible crack (usually excluding craze cracks). Crack sealing may also be called for in floors exposed to corrosive chemicals.

Some owners and designers insist on crack repairs solely for the sake of appearance, but that can backfire. The repair often ends up more obvious than the original crack.

Once you identify a crack as needing repair, you must decide how to do the job. Here are the four main methods:

- Fill the crack with a sealant.
- Fill the crack with a semi-rigid joint filler.
- Inject glue into the crack.
- Install dowels across the crack.

The choice of method depends not only on the nature of the crack, but also on how the floor is used.

Sealants

These are described in Chapter 15. The choices include hot-poured materials and cold-poured elastomers, which come in a bewildering variety.

Sealants make a good repair if the only requirement is that the crack be sealed. They do not protect a crack's edges from spalling under traffic, nor can they reduce differential movement.

Sealants require a minimum crack width, which varies according to the product. Most sealants need at least 3 mm (1/8 in). If the crack is not wide enough to admit the sealant, widen it with a concrete saw, a crack router or an angle grinder fitted with a mason's tuck-pointing blade.

Semi-rigid joint fillers

Based on either epoxy or polyurea, these are usually the best materials for cracks that have spalled, or are likely to spall, under hard-wheeled traffic. They even have some ability to stabilize cracks that show differential movement (Figure 13.4).

Because semi-rigids are less elastic than sealants, they will fail if cracks widen much after repair. To minimize the risk, wait as long as possible before filling cracks, to give the concrete time to shrink. And if the floor is subject to seasonal temperature swings, apply semi-rigid fillers during the coldest period, when thermal contraction is greatest.

Semi-rigids should penetrate at least 25 mm (1 in) into the slab, and may need to go deeper where the goal is to reduce differential movement.

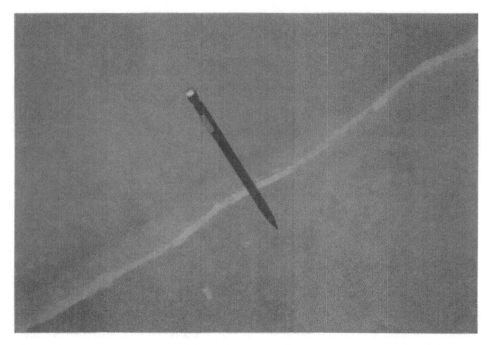

Figure 13.4 This crack was patched with semi-rigid epoxy. Not every crack patch looks this good, however

Unless the crack is unusually neat and clean, chase it with a concrete saw, a crack router or an angle grinder fitted with a mason's tuck-pointing blade. Make a groove at least 25 mm (1 in) deep and wide enough to take out all spalled edges (see Figure 13.5).

For the neatest repair, overfill the crack and sand or grind it flush after the filler has hardened. This works better with epoxy than with polyurea.

Glues

Some builders inject strong glue into cracks. The glue is usually epoxy (not semi-rigid) or methylmethacrylate.

Gluing is attractive because it is the only method that restores the floor's flexural stiffness to its uncracked condition. Another benefit is that is works in very narrow cracks; some products will go into a crack only 0.05 mm (0.002 in) wide.

Glue's strength is also its weakness. It locks up the crack, eliminating its ability to relieve future stress. If the conditions that caused the crack

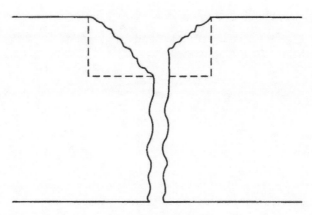

Figure 13.5 When repairing a crack with semi-rigid joint filler, always cut back to a clean, square edge

continue to exist, the floor is likely to crack again. Gluing is usually safe for structural cracks caused by temporary overloading, but not for drying-shrinkage and thermal-contraction cracks.

Dowelling

Some of the worst crack problems occur when cracks exposed to heavy industrial traffic lose their ability to transfer load. The two sides of the crack then move up and down relative to each other every time a vehicle passes over them. Once this starts, crack edges can break down quickly and severely.

If the differential movement is less than 0.5 mm (0.030 in), filling the crack with semi-rigid epoxy or polyurea may restore the lost load transfer. Failing that, the remedy is to install dowels, as described at the end of Chapter 14.

14

Curling

Curling is the warping of a concrete slab so its edges are left higher than its centre. The main cause is differential drying shrinkage through the thickness of the floor. If a concrete floor dries from the top – and ground-supported floors almost always do – the top shrinks faster than the bottom. This differential shrinkage warps the slab, making it concave upward (Figure 14.1). Thermal contraction can also curl a slab, but it plays a much smaller role than drying shrinkage – at least on interior floor slabs.

Curling is very common in ground-supported floors and concrete toppings, whether bonded or unbonded. It is a much smaller problem in suspended floors, because they are usually free to dry out from both top and bottom. Curling does show up, however, in slabs cast on metal deck.

A slab can curl up at any break – at the floor's edge, at a construction joint, at an induced joint or at a crack. The worst examples seem to occur at edges and construction joints, but some of that may be the result of false curling, described below. Severe curling also shows up at induced joints and cracks.

Curling starts very soon after construction and continues for many months. It has been suggested that over the very long term curled slabs settle back down due to creep. That may be true, but if so it takes years. The suggestion usually comes from someone who is trying to avoid repairing a severely curled floor.

The amount of curl varies greatly from floor to floor, for reasons that are only partly understood. In general, anything that increases drying shrinkage or increases the moisture differential will make curling worse. The list of contributing factors includes, but is not limited to:

- water content (Wetter mixes shrink more.);
- cement content (More cement means higher shrinkage.);

Figure 14.1 A kitchen sponge curls as it dries. So does a floor slab

- aggregate gradation (Concrete made with coarser, well graded aggregates shrinks less.);
- type of rock used for aggregate (Some types shrink less than others.)
- moisture content of sub-base (A wet base increases the moisture differential.);
- ambient humidity (Dry air increases the moisture differential.)

Contrary to common belief, curing seems to have little effect on how much a slab curls. Curing clearly affects when the timing of the process, since concrete will neither shrink nor curl while it remains soaking wet. But even the best-cured slabs dry out eventually, and will curl then.

Curling is so common we can hardly consider it a defect. We should rather think of it, like the drying shrinkage to which it is related, as a normal property of concrete floors.

But curling's frequent occurrence does not mean we should ignore it. Curling profoundly affects slab behavior, and severe curling can lead to these problems:

- debonding and break-up of concrete toppings;
- reduced structural capacity, due to loss of contact with the sub-base;
- differential movement at joints and cracks;
- reduced flatness at joints and cracks.

There are four general ways to deal with slab curl:
- Resist curling stresses with prestress or reinforcement.
- Limit curl by reducing drying shrinkage and moisture differentials.
- Design floors so curling does little harm.
- Repair curled floors.

False curling

Sometimes what looks like curling is really an upsweep at the slab edge, introduced during floor finishing. It results from settlement (see Chapter 7), which makes the fresh concrete surface lower than the side forms. When concrete finishers float and trowel a surface that has settled, they tend to pull mortar toward the edge to make the finished floor flush with the top of the side form or adjacent slab. This produces a short slope that can be as high as 6 mm (1/4 in). The length of the upsweep depends on the finishing tool used.

In their effect on flatness, false and true curling are almost the same. But unlike the real thing, false curling is strictly a surface defect. It does not reduce contact between a ground-supported floor and its sub-base, nor does it cause a topping to debond.

The most reliable way to tell real curling from false is to test the floor's flatness within a day or two of construction. Real curling never shows up that fast. If the time for early testing has past, there are still clues. Wherever curling appears confined to construction joints and is not evident at induced joints and cracks, suspect false curling.

Resisting curl

Prestress or reinforcement can resist and effectively prevent curling. Where, even a little curling would be too much – as in a random-traffic superflat floor – prestressing and heavy reinforcement are the only safe options.

Prestress

Floors prestressed with shrinkage-compensating concrete generally do not curl.

While the same cannot be said for post-tensioned floors, post-tensioning still prevents most curling-related problems because it eliminates most joints and cracks. A two-way post-tensioned slab curls only at its edges, which can be 100 m (300 ft) apart. In smaller buildings the post-tensioning can go wall-to-wall, putting any curled-up edges out where they are unlikely to cause trouble.

With one-way post-tensioning – common in defined-traffic superflat floors – you should expect normal curling in the direction perpendicular to the tendons. That is acceptable in defined-traffic floors where the critical traffic travels in just one direction.

Reinforcement

Reinforcing steel can resist curling, but you have to use enough. Structurally-reinforced slabs with double mats of steel almost never have problems related to curling. So-called jointless reinforced slabs, described in Chapter 16, have an equally good record for no or negligible curl.

On the other hand, the light reinforcement so often used in jointed slabs does not prevent curling. The best you can expect from it is some mitigation of the damages, since even light reinforcement improves load transfer across curled edges.

Limiting curl

The following steps help limit curl by reducing drying shrinkage:
- Use bigger coarse aggregate.
- Avoid aggregates that have high drying shrinkage.
- Use less water in the concrete mix (but beware the effect on workability and finishability).
- Use less cement (but beware the effect on workability and finishability).

Other measures limit curl not by reducing total shrinkage, but by reducing the moisture differential between top and bottom of the slab. The measures include keeping the air cool and moist above the slab, keeping the sub-base dry and deleting the damp-proof membrane.

The last measure, deleting the damp-proof membrane, is controversial. One school strongly opposes membranes, regarding them as one of the chief contributors to excessive curling. Members of this school delete membranes whenever they can. Where they cannot, they prefer to put a so-called blotter layer of sand or crushed-rock fines between the membrane

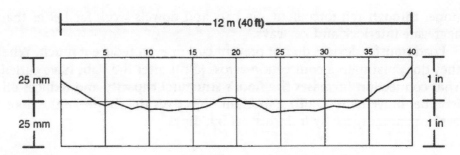

Figure 14.2 This slab curled not just from edge to edge, but also at the sawn joint in the middle

and the floor slab. The other school regards membranes as a minor factor, and points out that excessive curling occurs both with and without damp-proof membranes. The slab profiled in Figure 14.2 was laid over a granular sub-base, with no damp-proof membrane – and yet it curled badly.

Designing around curl

The methods that prevent curl cost money, and those that limit curl do not always produce the desired effect. Fortunately there is a third option: design the floor so curling does not matter. This is mainly a matter of getting the joints right.

Joint layout

Since curling problems show up mainly at joints, any design that keeps joints away from critical activities will improve the situation. In a warehouse or factory where curled joints affect the performance of wheeled vehicles, you can often put many of the joints in areas that get little or no traffic. Where concentrated loads control the floor design, careful layout can sometimes eliminate loading near joints.

Load transfer at joints

If you cannot keep the joints out of the critical areas, good load transfer will reduce many problems. While any load-transfer device is better than none, through reinforcement, tie bars and dowels work far better than aggregate interlock and keyways.

Load-transfer devices do not prevent curl or even reduce it much. What they do is maintain a connection across joints after the slabs have curled. That connection increases the floor's structural capacity and reduces differential movement – both important objectives that improve the floor's performance, especially in industrial buildings.

Repairing curled slabs

No slab should be repaired merely because it has curled. Repair is called for only if the curling creates a problem for the floor user. The repair method depends on the nature of the problem.

If the problem is confined to poor flatness or levelness, grinding is the usual repair. It works, but often has to go deeper and wider than people expect. Just running a grinder down the joint will seldom suffice. To eliminate the problem fully, the grinding swathe may have to extend as much as 1 m (3 ft) to each side of the joint. Diamond-disc grinders are best for this work; the grinders that use abrasive blocks are too slow.

Other curling-related problems may be harder to solve.

Once a thin topping has curled and begun to break up, it is beyond saving. At least part of it will have to be taken out and re-laid. If a thin topping has curled and debonded, but remains intact, its lifespan can sometimes be extended by injecting epoxy between the topping and its base. This is a tricky operation best carried out by specialists.

A similar operation, using cement grout instead of epoxy, is sometimes performed on ground-supported slabs that have curled away from their sub-bases. Grout injection usually restores support to the curled-up edge, but the repair does not always least. Even a tiny amount of further curling will lift the slab off the injected grout, recreating the problem.

If the problem is related to the loss of load transfer across a joint or crack, filling the gap with semi-rigid joint filler – epoxy or polyurea – sometimes saves the day. This repair works best where differential movement is less than about 0.5 mm (0.020 in). The semi-rigid material should fill the joint or crack as deeply as possible, and the surface should be ground flush after the material has hardened.

If joint filler alone does not restore load transfer – and in many cases it does not – the next tool in the kit is retro-fit dowels. The following steps describe one method:

1. Saw the floor at right angles to the joint or crack, making each sawcut 6 mm (1/4 in) wide, 100 mm (4 in) deep, 900 mm (36 in) long at the bottom and 300 mm (12 in) on centre.
2. Blow each sawcut clean with compressed air.
3. Put a steel flat bar, 6 mm (1/4 in) thick by 50 mm (2 in) deep by 900 mm (36 in) long in each sawcut, making sure the bar rests firmly on the bottom of the cut. Oil each bar to inhibit bonding.
4. Fill each sawcut with high-strength epoxy grout (not semi-rigid joint filler, which is flexible and weak for this task). Overfill the sawcut to leave the surface slightly crowned. Try not to let the grout enter the joint or crack.
5. Fill the joint or crack with a semi-rigid joint filler.
6. After both the high-strength epoxy grout and the semi-rigid joint filler have hardened, grind everything flush with the floor surface.

This repair is not suitable for slabs less than 150 mm (6 in) thick. Because it costs a lot, its use should be reserved for the high-traffic areas of industrial floors where differential movement causes real problems.

15

Joints

A joint is a planned break in the continuity of a concrete floor. An unplanned break is called a crack. The term *joint* is largely confined to those breaks that are more or less vertical, and not to the horizontal breaks between layers in a multi-course floor.

Joints are necessary but cause many problems. Close attention to the design and construction of joints will produce markedly better floors.

This chapter covers:

- the function of joints;
- types of joints;
- load transfer;
- how induced joints are made;
- fillers and sealants.

Chapter 16 deals with the related subject of joint spacing for crack control.

The function of joints

Floors have joints for two reasons:

- to make construction easier;
- to relieve stresses that would otherwise cause cracks.

To ease construction, all floors except the smallest are divided into several slabs cast on separate days. The seams between the slabs are called construction joints.

Many floors, but by no means all, contain stress-relief joints in addition to construction joints. Stress-relief joints include:

- isolation joints to separate the floor from other building elements;
- expansion joints to relieve stress from thermal expansion;
- contraction joints to relieve stress from drying shrinkage and thermal contraction;
- warping joints to relieve stress for curling.

A joint can serve more than one purpose. Many construction joints serve also as contraction and warping joints.

Ground-supported and suspended floors

Joints work differently in ground-supported and suspended floors.

Both kinds of floors contain construction joints, but the rules for locating them differ. In ground-supported floors, construction joints typically fall on column lines to make construction easier. In suspended floors, structural engineers prefer to put the construction joints in areas of low moment, which normally occur away from the structural supports.

A bigger difference shows up when we consider stress-relief joints. Most suspended-floor designs omit them altogether, relying instead on reinforcing or prestressing steel to resist stresses. In contrast, most ground-supported floor designs rely heavily on joints to relieve stress. Those joints are key parts of the floor design.

The rest of this chapter deals mainly with joints in ground-supported floors.

Joint types

Joints are classified in several ways: by function, by method of construction and by method of load transfer. Looking at function, we can divide joints into these categories:

- isolation joints;
- construction joints;
- expansion joints;
- contraction joints;
- warping joints.

Because a joint can do more than one thing, a single joint may fall in two or more categories.

Isolation joints

These separate a floor slab from other building elements, leaving the slab free to move as it responds to stresses from drying shrinkage, thermal contraction, loading and settlement. Isolation joints also protect the slab from movement that the other building elements undergo.

Isolation joints within a floor – that is, between floor slabs – are sometimes used where a floor must meet special requirements for thermal or acoustic insulation. Examples occur in cold-storage buildings and television studios. These look like and are made in much the same way as expansion joints.

Ideally, ground-supported floors are isolated from all objects that might restrict their movement, including walls, columns, machine foundations, drain pipes and bollards. Sometimes you have to compromise, however. Isolation is hardly possible where the structural design relies on a solid connection between slab and other building elements – a condition that occurs at many retaining walls and in portal-frame structures.

Isolation joints at walls are made by gluing or nailing joint filler to the wall. A 20-mm (3/4-in) thickness is standard in the UK, but 13-mm (1/2-in) joints are common in America and seem to work as well.

Isolation joints at columns are made in several ways (Figure 15.1). The simplest way is just to wrap the column with compressible foam, but this method is not suitable for every floor. A foam wrapping will not work where the column's support hardware sticks up above the sub-base. It also will not work on jobs where the columns go in after the slab.

The alternatives for column isolation are diamond-shaped or round blockouts. Diamond blockouts are usually formed up on site, using timber. Round blockouts are usually formed with special paperboard tubes, often known by the brand name Sonotube. In both cases, compressible joint filler goes between the slab and blockout. Blockouts are filled with concrete, a process then may either precede or follow the laying of the main floor slab.

Isolation joints are particularly important in post-tensioned floors, which experience large horizontal movements. Long post-tensioned floors need thicker isolation joints at building columns. If a post-tensioned slab is 100 m (330 ft long), each end may pull in by as much as 40 mm (1-$\frac{1}{2}$ in).

230   Design and Construction of Concrete Floors

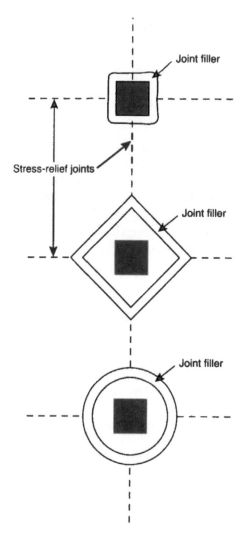

Figure 15.1 Three ways to isolate columns

Construction joints

These divide a floor into slabs cast at separate times. They are often called formed joints, because they are made by erecting forms. Another (somewhat dated) term for them is day joints, because they define each day's work.

Construction joints' main purpose is to simplify the job by dividing the floor into pours of convenient size. However, in many floors the construction joints also serve as stress-relief joints. Depending on details, a construction joint can serve as an expansion joint, contraction joint or warping joint.

Construction joints often cause trouble. The floor near a construction joint is almost always less flat than the floor elsewhere – sometimes much less flat. Recognizing that fact, American standards (ASTM E 1155 and ASTM E 1155M) exclude construction joints from routine flatness testing. Problems are not limited to poor flatness; construction joints often look bad and deteriorate quickly under traffic.

Good workmanship can go a long way toward preventing problems as construction joints. Still, the safest course is to eliminate or minimize construction joints in critical areas. Modern large-bay construction methods result in few construction joints, and this is perhaps the best argument in support of those methods. Where construction joints cannot be avoided, the following measures will improve their appearance and performance:

- Use side forms with sharp, square edges.
- When setting forms, make sure the inside edge (the edge nearest the pour) is higher than the outside edge.
- Compact the concrete well.
- When casting the second slab, keep the first one clean and do not let mortar build up on it.

Construction joints are finished in two ways. They can be tooled or trowelled flush (also called *burned in*). A tooled joint is made with an edging tool, which rounds off the slab edge. A flush joint is made by trowelling the second slab right up to the first. Tooled joints look better and are a good choice for exposed concrete floors that get light traffic. Flush joints may hold up better under industrial traffic, particularly where left unfilled. But flush joints often go wrong if they are sawn out later to receive a semi-rigid filler or sealant. The sawyer may find it hard to spot the exact location of the joint, resulting in a misplaced sawcut. In such a case, it's better to tool the joints using a tight-radius edger. The rounded-off edge should be narrow enough that the sawblade will remove it all.

If a construction joint ends up looking less than perfect, light grinding may improve it.

The important issue of load transfer at construction joints is covered later in this chapter.

Expansion Joints

These leave room for thermal expansion. Very few floors need them, because only in rare circumstances does a concrete floor expand beyond its original, as-cast size. Even in roads, which undergo greater temperature swings than do floor, modern practice eliminates expansion joints unless the concrete is cast in very cold weather. Some designers call for expansion joints in very big floors, and in long buildings where an expansion joint passes through the whole structure.

An expansion joint is made as a construction joint with a gap between the two concrete slabs. The gap is formed with joint filler, same as used in isolation joints.

Expansion joint do not fare well in areas of heavy industrial traffic, for two reasons. First, they are wider than most other joints. Second, their function demands that they be filled with soft, compressible material that has no ability to support the exposed joint edge. The best solution to this problem is to avoid expansion joints in trafficked areas. Another solution is to protect the joint edges with epoxy nosings, steel angles or steel flat bars.

If an expansion joint needs load transfer, steel dowels provide the only practical way (see Figure 15.2).

Contraction joints

These relieve tensile stresses caused by drying shrinkage and thermal contraction. They can be made either as construction joints (formed joints) or as induced joints.

All construction joints act naturally as contraction joints, unless restrained with reinforcement or tie bars. If a floor will include both construction and

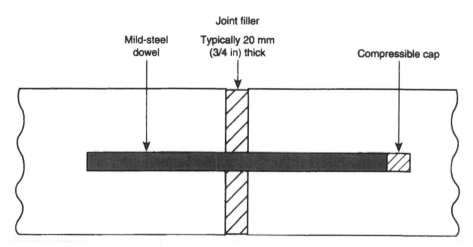

Figure 15.2 Dowelled expansion joint

contraction joints, it makes sense to combine the two functions where possible.

Induced joints are planned cracks. The slab is weakened – usually by sawing a groove, but sometimes by casting in a plastic strip – in the hope a crack will form along the weakened plane. Any induced joint will act as a contraction joint unless restrained by reinforcement or tie bars.

If a contraction joint needs load transfer, steel dowels provide the most reliable method. Some designers depend on keyways (at construction-contraction joints) or aggregate interlock (at induced contraction joints), but those methods are less reliable than dowels.

Warping joints

These act as horizontal hinges to relieve warping (curling) stresses. All expansion and contraction joints serve also as warping joints. Construction joints also relieve warping stress, unless they contain two or more layers of reinforcing steel.

When a joint is designed to work solely as warping joint, it usually contains tie bars or a single layer of reinforcing steel.

Load transfer at joints

Many joints need the ability to transfer vertical loads from one side to the other. Load transfer serves two quite different purposes. It increases a slab's structural capacity by reducing stress from edge and corner loading, and it improves the performance of joints under vehicular traffic.

To improve structural capacity, the load-transfer device needs to be strong but the joint need not be tight. A modest amount of differential movement can be accepted, provided the load-transfer device eventually takes up a substantial load. In contrast, joints exposed to vehicular traffic need to be tight, preventing all but the most minuscule differential movement. For best performance under industrial truck traffic, differential movement should not exceed 0.25 mm (0.010 in).

Here are the devices commonly used to transfer load across joints, ranked in order of generally decreasing effectiveness:

- prestressing steel through the joint;
- continuous reinforcement through the joint;
- tie bars;

- dowel bars;
- semi-rigid joint filler;
- aggregate interlock;
- keyways.

Some joints contain more than one type.

Prestress

Prestress across a joint provides near-perfect load transfer, whether the prestress comes from pre- or post-tensioning. This applies only to joints within the limits of the prestressed floor, however. Joints at the ends of a prestressed floor tend to open wide and often have problems with load transfer. Standard dowels may not suffice at the end of a very long post-tensioned floor slab, where joint opening may exceed 25 mm (1 in).

Through reinforcement

Reinforcing steel passing through a joint provides remarkably good load transfer. While the steel's shear strength plays a role here, the main benefit results from the way reinforcement keeps joints tight, maintaining concrete-to-concrete contact. It takes surprisingly little steel to do the job. Reinforcement equal to 0.12% of the slab's cross-sectional area has been used with success on some industrial floors. You can use too little, however. Very light reinforcement, such as the so-called driveway mesh used in America, may stretch so much that load transfer disappears.

Because through reinforcement limits joint opening (and closing), it cannot be used in isolation, expansion or contraction joints. It can be used in warping joints, however, provided the reinforcement is limited to a single layer.

Through reinforcement appears both in induced joints and in construction joints. Its use in induced joints creates no problems in construction, provided the steel lies far enough below the surface to allow sawing. Its use in construction joints complicates the job considerably, however, because the reinforcement has to pass through the formwork. One solution is to slot or split the forms, but that makes them less rigid and harder to maintain at the right elevation. Another solution is to use threaded bar connectors, but they add cost.

Tie bars

These are short lengths of reinforcing steel (deformed bar stock). Tie bars restrict joint opening, making them unsuitable for contraction and expansion joints. They allow hinging movement, however, making them ideal for warping joints.

Tie bars resemble dowels, but the two types work on different principles. Dowels are made of smooth bar stock so they can slip, letting the joint open. At a dowelled joint, the concrete faces may be several millimetres apart; vertical loads are transferred through the dowels. In contrast, tie bars are made of deformed bar stock to limit joint opening. The tie bars may transfer some load directly, but their main purpose is to keep the concrete faces close together.

While dowels have to resist bending and shear stresses, tie bars need only resist the tensile stresses that try to pull the joint open. For this reason, tie bars can be thinner and fewer than dowel bars. They need to be longer than dowels, however, to develop good bond with the concrete. In the UK, standard tie bars are 12 mm (1/2 in or #4) in diameter, 900 mm (36 in) long and spaced 600 mm (24 in) on centre. Deacon (1986) recommends reducing the spacing to 300 mm (12 in) where the floor supports concentrated loads that exceed 5 t (11 000 lb).

Tie bars work well with narrow-strip construction, though that is not as common as it used to be. Back when UK floors were often laid in strips 4.5 m (15 ft) wide, many designers put tie bars in most of the longitudinal joints, switching to dowels at every third or fourth joint. That pattern still works, producing a floor with many tight joints and a few wider ones. The tight joints, containing tied bars, serve as warping joints, while the wider joints, containing dowels, serve as contraction joints.

Not many American floors contain tie bars. But any American who wants to see them in action will not have far to look, since most concrete highways have them in the longitudinal joints, between the traffic lanes.

Dowel bars

These are smooth steel bars. They are usually round in section, but square-section dowels are preferred for some special cases. In recent years, flat-plate dowels have captured a growing share of the market. If installed correctly – a condition you should never take for granted – dowels transfer load even after a joint has opened up substantially, making them unique among load-transfer devices.

Dowelling has no rival as a way to transfer load across contraction and expansion joints. It is also used at warping joints, but there it faces competition from tie bars and through reinforcement.

Dowels find very wide use in construction joints, where they are easily installed by sticking them through the side forms. Their use in induced joints is less common, though no less effective. When installed at induced joints, dowels are supported on wire frames called baskets.

Dowel spacing and size

Dowels are usually spaced 30 mm (12 in) on centre at the slab's mid-depth. Closer spacing is rare. Wider spacing is sometimes seen, but should not exceed 600 mm (24 in) because a dowel's effectiveness goes down dramatically with distance from the applied load. To reduce restraint that can lead to cracks, some authorities recommend that no dowel be installed within 300 mm (12 in) of the point where two dowelled joints cross.

Dowel size is a matter of dispute, with opinions ranging all over the map. Table 15.1 shows the sizes used commonly (though far from universally) in America, and recommended by the American Concrete Institute (ACI 302). UK designers have traditionally specified much thinner (but no shorter) dowels for the same slab thickness. In a 150-mm (6-in) slab, where ACI 302 recommends 20-mm dowels, British practice would shrink the diameter to 12 mm – a 64% reduction in cross-sectional area.

The standard recommendations for dowel size are based mainly on tradition and experience, supported by some research performed in the fields of concrete roads and airport runways. Only rarely, have designers

Table 15.1 Dowel sizes (ACI 302)

Slab Thickness	Dowel Size and Spacing			
	Diameter	Cross-sectional Area	Length	Spacing
120–150 mm	20 mm	314 mm²	400 mm	300 mm
(5–6 in)	(3/4 in)	(0.44 in²)	(16 in)	(12 in)
180–200 mm	25 mm	491 mm²	460 mm	300 mm
(7–8 in)	(1 in)	(0.79 in²)	(18 in)	(12 in)
230–280 mm	35 mm	962 mm²	460 mm	300 mm
(9–11 in)	(1¼ in)	(1.23 in²)	(18 in)	(12 in)

tried to analyze specific load conditions, but that may be changing. The latest version of *TR 34* (Concrete Society, 2003, pp. 60-62) presents a way to calculate a dowel's load-transfer capacity. The method takes into account the amount of joint opening, a property ignored in other guides, but one that undeniably affects a dowel's efficiency.

Allowing slippage

To do its job, a dowel must be free to slip in the direction perpendicular to the joint. At the same time, it must be closely connected to the surrounding concrete so it can transfer load across the joint.

Three things are necessary for slippage. The dowel must be straight; it must be perpendicular to the joint, and it must not bond to the concrete. Failure to meet any of those conditions turns a dowel into a tie bar, making it unsuitable for a contraction or expansion joint.

Ordinary bar stock is straight enough for dowel use, but cutting it the wrong way can ruin it. Always insist that dowels be sawn to length, not sheared. Shearing costs less but puts a hook in the dowel's end, and that should never be accepted.

Keeping the dowel perpendicular to the joint is the part that most often goes wrong. At construction joints, many floorlayers like to leave the dowels loose during the pour, straightening them by hand after the concrete has stiffened slightly. That can work, but it depends on close attention at exactly the right time. Safer if more costly method is to support the dowels rigidly before placing concrete. Plastic mouldings called dowel aligners are available for that purpose. They work best with timber forms, to which they can be nailed. Another option, suitable for any type form, is to erect a frame outside the form to support the dowels.

Despite the undoubted importance of keeping dowels aligned, standard specifications say little about what tolerances are needed. Deacon (1986) recommends that the deviation from perpendicular not exceed 6 mm in 300 mm (1/4 in 1 ft).

Preventing bond is a matter of putting something between the steel dowel and the surrounding concrete. That something can be grease, heavy oil or a plastic sleeve. Plastic sleeves come in thin and thick varieties, bearing different risks. The thin sleeves are made of thin, flexible, clear polyethylene. The risk with them is that they will be dislodged during the concrete pour, since they are light and flimsy. The thick sleeves are made of rigid plastic, and the main risk here is that they will not fit tightly over the dowel. Because the thick sleeves are more easily compressed than steel or concrete, some people worry that they may yield too much under load, resulting in a joint that moves under traffic. Some of the thick plastic sleeves are meant to fit over deformed bar stock, allowing the use of ordinary reinforcing bars as dowels.

To save money and effort, debonding is usually confined to one side of the joint, though it does no harm to debond both sides. When dowels are installed at induced joints, it is good practice to debond more than half each dowel's length, to allow for the possibility the joint will not be perfectly centred on the dowels.

One extra detail is needed where dowels are used in an expansion joint. There each dowel needs a compressible cap at one end, and that end must be debonded. The cap should be at least as thick as the joint is wide.

Occasionally, dowels must be installed where a new floor abuts an old one. The usual solution is to drill holes in the old slab, glue the dowels in the holes with a strong epoxy, and debond the dowels from the new concrete with grease, oil or sleeves.

Non-traditional dowel shapes

While round-section bars continue to dominate, other shapes are gaining in popularity. Some designers favour square-section bars (see Figure 15.3). If padded along its vertical faces, a square dowel allows some horizontal slippage parallel to the joint, while still transferring vertical loads. This can be useful where the slabs on either side of the joint are likely to shrink at different rates or in different directions, as might occur where a post-tensioned slab abuts one without post-tensioning.

A newer development is the plate dowel, made of steel plate typically 6–10 mm (1/4–3/8 in) thick. At construction joints, you use square plates oriented diamond-fashion, so the joint falls across two opposite corners. Installation involves nailing a hard plastic pocket former to the side form. After casting the first slab, you strip the form and the nails come away with it, leaving the pocket former embedded. You then insert the plate dowel before pouring the second slab.

Plate dowels can also be used at induced joints, though that is less common. The pocket formers won't work there. Instead, you use rectangular flat bars on wire frames. A plastic sleeve covers just over half the length of each dowel.

Plate dowels seem to use steel more efficiently than the tradition round bars, though that has not necessarily translated into a lower price. The drawbacks are difficult in compacting the concrete under the plate, and the risk of a loose fit between dowel and pocket former.

Joints

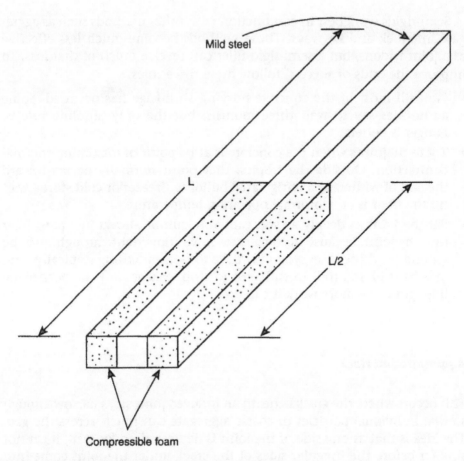

Figure 15.3 Square dowel with padded vertical faces

Semi-rigid joint filler

Based on epoxy or polyurea, these were developed to help joints withstand wear and tear from warehouse trucks. They have proven effective in improving load transfer under certain conditions. They work by eliminating the gap at the joint.

We should not expect any semi-rigid to serve on its own as the sole load-transfer device. For example, at a plain butt construction joint 6 mm (1/4 in) wide, no amount of semi-rigid filler is likely to provide any effective load transfer.

Semi-rigids work best in conjunction with other methods such as aggregate interlock and keyways. Those methods become much less effective as a joint widens, but a semi-rigid filler can reverse much of that loss. To improve the odds of success, follow these three rules:

- Wait till most of the concrete's drying shrinkage has occurred. Some authorities say to wait three months, but the only absolute rule is: Longer is better.
- Try to fill joints when the concrete is at its point of maximum thermal contraction. Outside the tropics, that point normally occurs toward the end of winter. In a refrigerated building (freezer or cold store), wait till the floor is at its normal operating temperature.
- Fill the joint as deeply as you can. You cannot always just pour filler into the joint, because in some cases it will flow right through into the ground. In those cases, you may have to put something (typically sand or a hard plastic rod) in the joint to stop leakage. But the deeper the filler goes, the more work it can do.

Aggregate interlock

This occurs where the crack beneath an induced joint stays narrow enough so that individual particles of coarse aggregate can touch across the gap. The idea is that as one side of the joint is depressed under load, it cannot sink far before the irregular sides of the crack under the joint come into contact, transferring load.

As I say, that is the idea. The reality is that aggregate interlock loses effectiveness dramatically as the crack beneath the joint widens, and becomes useless after the width reaches 1 mm (1/32 in) or so. ACI 302 warns us not to rely on aggregate interlock where crack width exceeds 0.9 mm (0.035 in). That is not encouraging advice for those who might hope to use this method as the sole method for load transfer. If an unreinforced floor has joints 4 m (13 ft) apart, and if the concrete shrinkage amounts to 0.050%, and if all joints activate, then we can expect a mean crack width of 2.0 mm (0.078 in) – over twice the width at which ACI 302 says aggregate interlock ceases to be reliable.

Unless the joint spacing is absurdly close – 2 m (6 ft) or less – aggregate interlock is of dubious value in a floor that lacks steel at the joints. On the other hand, it plays an important and useful role in floors where through reinforcement or steel fibres keep joints narrow.

It has been suggested, plausibly, that coarser and elongated aggregates enhance aggregate interlock.

Aggregate interlock applies only at induced joints (and cracks). It does not exist at construction joints.

Keyways

These are tongue-and-groove joints (see Figure 15.4). Like aggregate interlock, keyways rely on concrete-to-concrete contact, and lose effectiveness as the joint opens. Unlike aggregate interlock that applies only to induced joints (and cracks), keyways are installed in construction joints.

A specially-shaped side form makes the key. Builders who use timber forms generally choose wood for the key former as well, nailing it on. Builders who prefer steel forms can find them with key formers already attached.

Inducing joints

Construction joints are made by casting concrete to edge forms that run the full slab depth. All other joints, whatever their function, are induced.

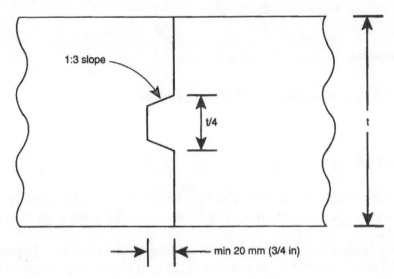

Figure 15.4 Keyway

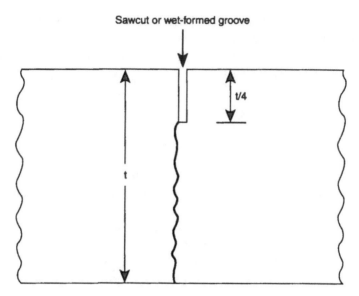

Figure 15.5 Induced joint

An induced joint is really a planned crack (Figure 15.5). The floor slab is notched or cut to create a weakened plane, in hope that a crack will occur there rather than elsewhere. The notch can be made by wet forming or sawing. Wet-formed joints costs less, but sawn joints hold up better under traffic and are easier to fill.

Wet forming

The choices here include hand tooling and plastic inserts.

Hand tooling

With this traditional method, the floorlayer uses a jointing tool to make a groove in the wet concrete. The tool leaves the joint's edges slightly rounded.

Hand tooling can produce a good-looking joint, but it is only practical for thin slabs because it goes, at most, 25 mm (1 in) deep. Most authorities say the joint depth should be at least one fourth the slab thickness, limiting hand-tooled joints to slabs not more than 100 mm (4 in) thick.

Because hand-tooled joints have rounded edges, they should not be used in areas of heavy traffic. At first thought, it might seem that rounded edges would hold up better than square ones, but the reverse is true. Under heavy traffic any induced joint is likely to need filling, and it is hard to do a good job of filling a joint with rounded edges. The issue is usually moot, because most floors that get heavy traffic are too thick for hand tooling.

Plastic inserts

These are strips of plastic that are pressed down into the wet concrete and left there. Many people call them Zip Strips, after the best-known brand. Some types include a top strip that is peeled away later to leave a groove for sealant.

Unlike hand tooling, plastic strips will work in thick slabs. Even so, they are seldom the best choice for floors that get heavy traffic. Sawing, described below, generally produces a more durable joint.

Never use plastic inserts in superflat floors, because it is hard to make a floor flat near an insert.

Inserts can be hard to install. If pushed in with too much force, they sometimes bend to the side, resulting in a wedge-shaped sliver of concrete that easily breaks up. It helps to run a trowel, with blade held vertical, along the line of the joint. That clears away the coarse aggregate to leave a path for the insert.

Sawing

This costs more than wet forming but makes neater, more durable joints for industrial floors. Sawing is a good choice for floors that need to be flat, because it has almost no effect on surface regularity.

Joint-cutting saws fall into two categories: conventional and early-entry. Each type employs a circular blade driven by a small petrol engine or an electric motor.

Conventional saws have silicon-carbide or diamond blades and are normally used with water. They can be used dry but then raise huge amounts of dust. Most models have a downcutting blade; that is, cutting takes place as the edge is travelling down into the slab.

Early-entry saws (Soff-Cut is the best-known brand) have diamond blades and are used dry. A flat bearing plate slides along the floor surface;

its purpose is to hold the concrete in place as the upcutting blade slices through. The bearing plate allows an early-entry saw to be used on soft, weak concrete.

Early-entry sawing grew in popularity throughout the 1990s, but is not without its critics. Some people believe that very early sawing, made possible and encouraged by the introduction of the early-entry saw, leads to weakened joints that break down prematurely under industrial traffic. This is a question that cries out for research.

Decisions about sawing centre on two issues: depth and timing. The two are somewhat related.

Sawcut depth

This is always a compromise. A deeper cut increases the likelihood that slab will crack beneath it, activating the joint. On the other hand, a shallower cut costs less, goes faster and leaves a thicker section below to provide aggregate interlock, possibly improving load transfer.

Traditional rules commonly called for sawing to one fourth the slab depth, with more conservative designers increasing that to one third. ACI 302 recommends one third when the concrete contains steel fibres, presumably because they inhibit crack formation beneath the sawcut.

Those rules have been relaxed in recent years with the introduction of the early-entry saw, which allows earlier sawing than is possible with conventional saws. The first version only cut about 30 mm (1-1/8 in) deep. The manufacturer claimed that was enough to force the desired crack, provided the sawing took place early. More recent versions cut deeper, but even now designers often allow shallower joints if an early-entry saw is used.

Timing

When should you saw joints? As with sawcut depth, the answer always involves a compromise. Saw too early and you get a rough joint vulnerable to damage. Wait too long and the slab will have cracked. Some designers specify a time limit, but that can never replace careful observation and good judgment on the sawyer's part.

The general rule holds that sawing should occur as soon as the concrete can stand it without crumbling. The exact timing depends on concrete mix and type of saw, and – most of all – on temperature. In the hottest weather, you can start sawing a few minutes after the last trowel pass. In very cool weather, you may have to wait till the next day, though that always brings the risk that overnight cooling will cause thermal-contraction cracks.

Early-entry saws, as their name implies, allow sawing to start earlier – though not as much earlier as you might suspect. In warm weather, there is not much difference in starting time between early-entry and conventional saws, but the spread widens in cool weather. While it is never good practice to wait longer than necessary to saw joints, early-entry saws introduce another reason to hurry: they can create dust. When the cutting takes place early, moisture in the concrete keeps the dust down. But if you wait even an hour or two, you can get annoying amounts of airborne dust. The problem is most acute on big pours where the sawing does not keep pace with the concrete's drying.

Joint fillers

The term covers two quite different kinds of products:
- sheet fillers used in expansion and isolation joints;
- semi-rigid joint fillers.

Sheet fillers

These are solid, compressible materials used in expansion and isolation joints. A sheet filler has two functions. It defines the width of the joint, and it keeps out debris. It is not meant to seal the joint against liquids.

Sheet fillers must be installed before the concrete is cast. They are almost impossible to install afterwards.

The most common type is called filler board, made of vegetable fibre bound with bitumen or asphalt. Another type consists of bituminous mastic sandwiched between two layers of thin felt. A third type is made of plastic foam, often coloured grey to match the concrete.

Sheet fillers must be compressible. Compressibility can be defined by the pressure needed to squeeze the material down to half its original thickness. According to American standards, vegetable-fibre (ASTM D 1751) and bitumen-sandwich (ASTM D 994) filler boards should test between 0.7 and 5.2 MPa (100 and 750 psi). For plastic foam (ASTM D 1752) the limits are wider – 0.3 and 10.4 MPa (50 and 1500 psi).

Those compressibility limits are probably good enough for expansion joints, where the effect of stiffer material is to increase compressive stress in the concrete, which can easily resist such stress. The story is different at certain isolation joints, where stiffer material can increase tensile stress in the concrete, which may then crack. The solution is to make sure the compressibility lies near the soft end of the allowed range. Good results have been obtained with sheet fillers that require no more than 1 MPa (150 psi) to compress them by half.

Filler board comes in various thicknesses, normally chosen to match the specified joint width. For very wide joints, as found in post-tensioned floors, you may have to use multiple layers.

A common mistake is not to install filler board the full depth of the slab. If the slab is designed to be, say 150 mm (6 in) thick, most builders will buy filler board 150 mm (6 in) wide, or cut the board to that width. That seems right, but it presumes the slab will not exceed the specified thickness. As we saw in Chapter 5, slab thickness varies more than you might expect. If our so-called 150-mm (6-in) slab happens to be 180 mm (7 in) thick where it abuts a building column – by no means a rare occurrence – then a 150-mm (6-in) strip of filler board will leave a substantial length of bare column sticking out, defeating the purpose of the isolation joint.

Sheet fillers keep debris out of joints, but beyond that they do not substitute for sealants. If an expansion or isolation joint needs to be sealed against liquids, the top of the filler should be cut away and the gap filled with a high-movement elastomeric sealant. Some joint fillers are scored to make this step easier.

Semi-rigid joint fillers

These help protect joints from damage caused by hard-tyred industrial vehicles. They start out liquid and are poured or injected with a caulking gun into the joint. They harden into a firm but slightly flexible solid – hence the term *semi-rigid* (Figure 15.6). After curing, they typically have a Shore A hardness of about 80.

Semi-rigids resemble joint sealants but differ from them in important ways. Semi-rigids protect the joint edges from mechanical damage; sealants keep liquids out of the joint. Semi-rigids are harder and stronger than sealants, but much less elastic.

Semi-rigids stick to concrete, but not tenaciously. One typical product has a reported tensile strength of 3.1 MPa (450 psi), but a bond strength

Figure 15.6 Warehouse floor joints filled with semi-rigid epoxy

to concrete of only 1.4 MPa (200 psi). This is deliberate, to ensure the filler does not glue the joint so tightly that the slab cracks elsewhere.

Semi-rigid fillers are made of epoxy or polyurea resins. Both materials have their supporters, and the jury is still out on which does the best job. Epoxy, has a longer history, is generally a little harder, and is easier to install without special equipment. Polyurea cures much faster.

Both products are costly, but you can minimize cost by making joints narrow and filling only those joints that will be exposed to traffic. It is a waste to use any semi-rigid where it will be covered by racks or other fixed equipment.

The biggest drawback to semi-rigids is that they accommodate only very small joint movements. That is the price we pay for their strength and hardness. An ideal joint filler would be both strong and highly elastic, but that product does not exist and may not even be a possibility. If you install a semi-rigid soon after laying a floor, the risk is high that some joints will re-open from drying shrinkage (see Figure 15.7).

The imperfect solution to that problem is to wait as long as possible before filling joints. Some specifications require a 90-day delay, but that's arbitrary. You cannot wait too long, provided you fill the joints before they start to deteriorate from traffic.

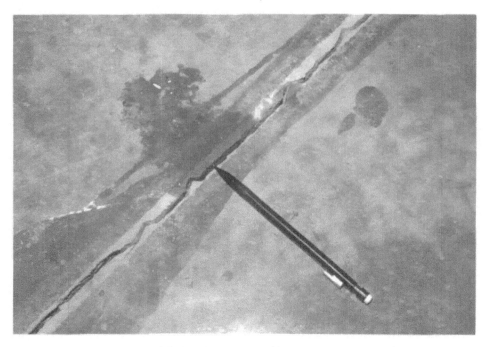

Figure 15.7 This joint was filled with semi-rigid filler, but re-opened as the concrete shrank

Waiting only works where the joint movement is caused by drying shrinkage, which diminishes over time. If joints are subject to large cyclic movements – as in an unheated warehouse built where winters are cold – semi-rigids are not very effective.

Poor installation often keeps semi-rigids from living up to their potential. To be fully effective, these fillers need to go deep and be finished flush with the floor surface. At sawn joints, the filler should go all the way to the bottom of the kerf. At construction joints, it's good practice to chase the joint with a saw, cutting at least 25 mm (1 in) deep, thus creating a shoulder for the filler to bear on. The most reliable way to ensure a flush joint is to overfill, wait for the material to harden and then sand or grind off the excess. That works better with epoxy than with polyurea. The alternative, not quite so reliable, is to overfill and shave off the excess with a blade while the material is still soft.

A dilemma occurs when a joint needs both a semi-rigid filler to protect against wheeled vehicles and a sealant to keep out liquids. This condition occurs often in food and pharmaceutical factories and warehouses. Semi-rigid fillers are too stiff to make good sealants, while elastomeric sealants (described below) are too soft to protect joints from traffic. There is no practical way to use both products in the same joint, so we have to choose one and accept that it won't do the other's job.

One way around that dilemma is to get rid of the joints. Floors with post-tensioning, shrinkage-compensating concrete or heavy continuous reinforcement need very few joints, but usually cost more.

Armoured joints

Joint deterioration under heavy traffic is a big problem in warehouses and factories. Semi-rigid joint fillers reduce the problem but do not eliminate it.

An alternative is to gird the joint's edges with steel. Steel angles or steel flat bars (on edge) are cast into the concrete on each side of the joint. Sheer studs connect the steel to the concrete. The result is that the joint edges consist not of concrete but of steel, which can better resist impact and wear from vehicles.

Regrettably, the steel-armoured joint has not always lived up to its promise. It is one of those details that always looks good on the engineer's drawings, but is easily wrecked by poor workmanship on site. For good results, the steel bar or angle must be perfectly flush with the adjacent concrete, as well as level with the bar or angle on the other side of the joint. It is probably easier to get a good job with flat bars than with angles, but failures have occurred with both shapes. Figure 15.8 shows an armoured joint that needed extensive grinding.

Joint sealants

People seal joints for these reasons:
- to protect floors from chemical attack;
- to keep debris out of joints;
- to make buildings easier to clean;
- to make floors look better.

Not every joint needs to be sealed. There is seldom any reason to seal joints in slabs that will get floorcoverings. Even where the concrete slab forms the wearing surface, joint sealants are often unnecessary. On the other hand, some floor users – notably in the food and pharmaceutical industries – insist on the careful sealing of every joint.

250   Design and Construction of Concrete Floors

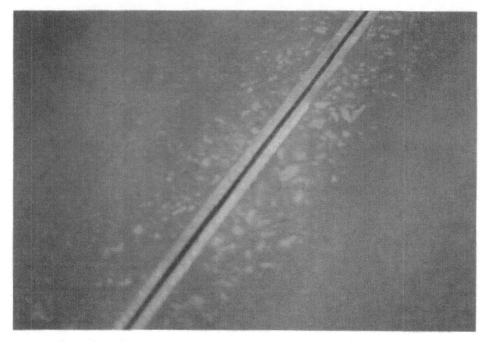

Figure 15.8 This armoured joint (flat-bar type) needed grinding

Depending on the situation, a joint sealant may need any of these properties:

- low cost;
- long life;
- elasticity to accommodate joint movement;
- resistance to chemical attack;
- paintability.

Because no material possesses all those properties, designers must decide which properties matter most for each floor.

Complicating the choice is the fact that this field is full of proprietary products, some covered sketchily or not at all by national standards. Many designers specify sealant by brand name, because that is simpler than specifying by type or performance. For those up to the challenge of writing more open specifications, ACI 504, BS 6213 and ISO 11 600 offer some help.

Manufacturers' catalogs describe sealants in many ways, but most materials suited for floor joints fall into one of these two categories:

- hot-poured sealants;
- elastomeric sealants.

Hot-poured sealants

A very old way to seal joints is to fill them with hot, molten bitumen and pitch. Hot-poured sealants are still in use, though cold-poured elastomers have replaced them for many applications.

Hot-poured sealants are cheap but have these shortcomings:

- They cannot be applied in narrow joints.
- They do not last long, having a typical lifespan of 3 to 10 years.
- They accommodate only medium joint movement – up to about ±12.5% of the joint width.
- They are attacked by some common solvents.
- Coatings will not stick to them.

Elastomeric sealants

These find use in isolation, expansion and contraction joints where joints are expected to open and close. High-movement elastomers can accommodate movement greater than ±12.5%. Medium-movement elastomers are designed to expand and contract by up to 12.5%. (If there are any low-movement elastomers, no one says much about them.)

Most elastomers are described as cold-poured. This means they are installed at room temperature, in contrast to hot-poured sealants that heated to make them flow. Some harden by chemical reaction, while others just dry out.

The category includes widely different chemical types, some of which come in two parts that are mixed just before application. Two-part sealants include polyurethanes, epoxies and some polysulphides. One-part sealants include silicones, acrylics and some polysulphides. Many come in both high-movement and medium-movement formulations.

Installing joint sealants

Different products require different techniques, and there is no substitute for reading the manufacturers' instructions. Nevertheless, a few general principles apply.

The joint should be clean. In some cases, cleaning is just a matter of scraping out debris or vacuuming out the joint. If the joint is very dirty, cleaning may require chasing with a concrete saw or crack router.

The joint must be wide enough. It is hard to install any sealant in a joint less than 3 mm (1/8 in) wide, and some materials need greater widths. Some construction joints need to be sawn open before they can accept a sealant. (But a very tight construction joint may not need to be sealed at all.) Wider joints can accommodate more movement before the sealant fails.

The sealant should be installed to the right depth. In contrast to semi-rigid fillers, which go down at least 25 mm (1 in), most elastomeric sealants work best if their depth is about equal to their width. Some one-part elastomers have to be installed in thin layers in order to cure properly.

16

Crack Control in Ground-Supported Floors

Concrete floors crack in many ways (see Chapter 13), but in ground-supported floors the most common and most troublesome cracks are those caused by drying shrinkage and thermal contraction. In this chapter, the term *crack* refers only to that type.

We are talking here about crack control in ground-supported floors. Suspended floors crack, too, but the structural codes dictate that those floors contain enough reinforcing or prestressing steel to keep cracks from becoming a serious problem. In contrast, most ground-supported floors are not covered by the codes and sometimes contain no steel at all.

Designers have three tools for controlling cracks: joints, reinforcement and prestress. Joints create planned breaks in the floor, in contrast with the unplanned breaks we call cracks. Reinforcement does not prevent cracks, but it limits their width. Prestress prevents cracks by putting the concrete into compression.

Each of those three tools can be used alone. In addition, the first two – joints and reinforcement – can be used together. Thus we have four possible solutions:

- an unreinforced floor with joints;
- a reinforced floor without joints;
- a reinforced floor with joints;
- a prestressed floor.

The "let it crack" approach

Before we get into the details of crack control, it's worth noting that a few people do not believe in it. They design unreinforced or nominally reinforced floors with no control joints, and they let the concrete crack where it will. The argument is that the cracks that occur naturally are less numerous and no more annoying than the joints that would have been necessary to prevent them.

No one seriously recommends this approach for exposed industrial floors, but it has been used successfully for slabs that receive floorcoverings and foot traffic, including some floors for big retail stores.

Unreinforced floors with joints

These are cheap and easy to build – particularly if the usage does not demand load transfer or filling at the joints.

Unreinforced floors have never been numerous in the UK, though they do exist there. They are far more popular in America, where the widespread use of laser screeding has encouraged the design of floors that need no reinforcing steel. Laser screeding does not require unreinforced floors, but it goes faster and costs less with no embedded steel in the way.

Joint layout

Joints are normally laid out in a grid, crossing the floor in two perpendicular directions. Because there is no steel to control random cracks, joints must be spaced closely to make such cracks unlikely.

But what is the safe distance between joints? That is one of the biggest questions in floor design. Everybody has an opinion, and the opinions vary widely. It seems unlikely that any firm answer will ever arise, because it depends on concrete shrinkage, site condition and – perhaps most of all – on the floor user's tolerance for the odd random crack. The truth is that no joint spacing can totally guarantee a crack-free floor, and it comes down to probabilities. If we shorten the distance between joints, the probability of cracks goes down – always. But it never goes down to zero.

An old rule of thumb, still widely used, holds that the joint spacing measured in feet should not exceed two or three times the slab thickness measured in inches. To put it in terms that do not depend on feet and inches, the joint spacing should not exceed 24 to 36 times the slab thickness.

For a 150-mm (6-in) slab, the safe maximum joint spacing would range from 3.6 to 5.4 m (12 to 18 ft). Spacings near the lower end would apply where shrinkage is not well controlled (see Table 16.1).

The current version of ACI 302 keeps the old standard rule tying joint spacing to thickness, but imposes an overriding maximum spacing of 5.5 m (18 ft) no matter how thick the slab. However, the next version will probably drop the rule and recommend closer spacings.

Recent American practice, with or without official guidance, has been shifting toward closer joint spacings and less attention paid to any connection with slab thickness. Many American designers routinely limit joint spacing to about 4.0 m (13 ft), and a few reduce that to 3 m (10 ft).

The argument in favour of closer joint spacings goes beyond crack control. A close spacing reduces the effects of slab curl, and reduces the opening at each joint. On the other hand, all those joints cost something to make, and in an industrial floor can be costly to maintain.

Joint spacing should be about the same in both directions. In no case should any slab panel (bounded by joints) have an aspect ratio (length divided by width) over 1.5. Square panels work best.

Many situations require closer joint spacing than the theoretical maximum. Any big slab penetration such as a building column acts as a crack inducer, even if isolated. In buildings with interior columns, joints should be located on all the column lines in both directions, and this requirement often controls the joint spacing, which needs to be a fraction – typically 1/2, 1/3 or 1/4 – of the column spacing.

Most designers do not add joints at minor penetrations such as drain cleanouts. But slabs sometimes crack there.

Table 16.1 Maximum joint spacing in unreinforced slabs (Portland Cement Association)

Slab Thickness		Maximum Joint Spacing					
		Slump Above 100 mm (4 in)				Slump Below 100 mm (4 in)	
		Aggregate Under 20 mm		Any Aggregate Over 20 mm			
mm	in	m	ft	m	ft	m	ft
125	5	3.0	10	3.8	13	4.5	15
150	6	3.6	12	4.5	15	5.4	18
175	7	4.2	14	5.2	18	6.3	21
200	8	4.8	16	6.0	20	7.2	24
225	9	5.4	18	6.8	23	8.1	27
250	10	6.0	20	7.5	25	9.0	30

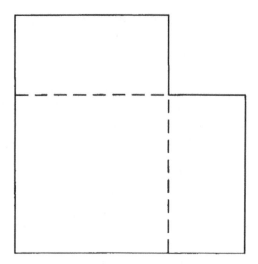

Figure 16.1 Joints along the broken lines would reduce the risk of a crack at this re-entrant corner

It is always a good idea to provide extra joints at re-entrant corners (Figure 16.1), which are notorious crack-inducers. Some designers add reinforcing steel at such corners even if the rest of the floor lacks it.

Types of joints

The crack-control joints in an unreinforced floor can either be construction joints (formed joints) or induced joints (usually sawcuts). Many floors contain both types.

In chequerboard construction, rare these days, all joints are construction joints. In narrow-strip construction, all the joints running one way are construction joints, and all running the other way are induced. In wide-strip or large-bay construction, most joints in both directions are induced.

Load transfer at joints

Some unreinforced floors – mainly those in warehouses and factories – need the ability to transfer load across joints.

The construction joints pose no special problems. The usual devices for transferring load across construction joints work the same in unreinforced slabs as in other slab types. Round dowels, square dowels, plate dowels and keyways have all been used.

Induced joints are another matter. Here we face the choice between relying solely on aggregate interlock and adding dowels on wire frames. Aggregate interlock comes free but is not very reliable. It works better where joint spacing is close, shrinkage low and aggregates big. Dowels perform more reliably but complicate the job. If the dowels are installed crooked, they can lock up the joints, causing cracks between the joints. Even straight dowels increase the cracking risk somewhat. Some designers dowel the joints in traffic areas only.

Benefits and drawbacks

Unreinforced floors offer these benefits:

- low initial cost;
- easy access for concrete trucks and laser screeds;
- reduced risk of plastic-settlement cracks;
- no risk of steel interfering with wire-guided vehicles.

They also have drawbacks:

- Cracks, if they occur, can open wide.
- Many joints can create a long-term maintenance problem.

In general, unreinforced floors make the most sense where traffic is light and joints do not need much load-transfer ability. They are harder to justify – though some people still do – for heavily-trafficked industrial floors.

There is a strong prejudice in some quarters against unreinforced floors, even in buildings that get only foot traffic, but the evidence does not seem to support that prejudice. Many successful floors contain not a shred of reinforcing steel. Some failures have occurred, too, but that statement can be made about every crack-control method. It is worth noting that unreinforced concrete slabs, usually with dowelled contraction joints, have been installed in thousands of miles of streets and roads.

Reinforced floors without joints

Though not common, these have an excellent record of success. They rely on substantial quantities of reinforcing steel to keep cracks narrow. A reinforced jointless floor may have many cracks – perhaps as many as one per metre (one per yard), but the cracks cause little trouble as long as the steel keeps them tight.

These floors are also known as continuously-reinforced concrete pavements (CRCPs), a term that comes from the highway field.

Two-way and one-way jointless floors

These floors can be joint-free in one or two directions.

Two-way jointless floors are heavily reinforced in two perpendicular directions. They contain no joints except the usual isolation joints at columns and walls, and construction joints at the end of each day's pour.

One-way floors are heavily reinforced in one direction only (see Figure 16.2). Control joints run parallel to the reinforcement, dividing the floor into long, narrow strips. One-way floors rely on different methods of crack control in the longitudinal and transverse directions. Longitudinally, each strip is designed as a reinforced jointless slab and contains a substantial amount of longitudinal steel. Transversely, each strip is designed as an

Figure 16.2 One-way continuous reinforcement in a defined-traffic superflat floor

unreinforced jointed slab and has little or no transverse steel – typically just enough bars to support the heavy longitudinal steel.

How much steel?

The reinforcement is usually stated as a percentage of the concrete's transverse cross-sectional area, using this equation:

$r = 100 A_r(td)$

where:

r = reinforcement, in %
A_r = cross-sectional area of reinforcing bar, in mm² or in²
t = slab thickness, in mm or in
d = bar spacing, centre to centre, in mm or in

The closest thing we have to a standard amount of steel for jointless construction is 0.5%, in one or two directions. Since that percentage is based on experience rather than calculations, I want to explain where it comes from.

Long reinforced slabs with no transverse joints are common in roadbuilding, where almost a half century of experience has taught highway engineers how much steel they need. The Federal Highway Administration (1990) recommends at least 0.6%, based on the use of Grade 60 deformed bars. American Grade 60 steel has a characteristic strength of 414 MPa (60 000 psi), and the nearest UK equivalent is British Grade 460. The Federal Highway Administration suggests that a lower percentage of steel may be acceptable "where experience has shown that [it] has performed satisfactorily". A higher percentage, 0.7%, is recommended only in places with extreme winter cold (mean minimum monthly temperature below −12°C (10°F).

It seems, then, that 0.6% reinforcement might be a reasonable specification for a jointless concrete floor. But what's appropriate for a highway pavement may be more than a floor really needs. Floors are shorter than roads and are subject to less extreme temperature swings. With those mitigating factors in mind, some floor designers in the 1980s chose, somewhat arbitrarily, 0.5%. Since then many floors have been built with 0.5% reinforcement in one or both directions, and none has been known to fail in a way that suggested the amount of steel was not enough. That is the reason behind the current use of 0.5% steel in jointless floors. Some people have speculated that even less than 0.5% might suffice, especially for short slabs, but the research to support that remains undone.

Two-way jointless floors take 0.5% reinforcement in each direction. One-way jointless floors need the 0.5% in just one direction. They could be built with no reinforcement at all in the other direction, but typically have a few cross bars to support the longitudinal bars. Typical transverse reinforcement consists of 12-mm (1/2-in or #4) bars spaced 1.2 m (4 ft) on centre. Some designers put in more, following the principles of reinforced slabs with joints.

Reinforcement details

The reinforcement usually consists of individual deformed bars, made of British Grade 460 or American Grade 60 steel. Welded mats have also been used.

Bar spacing varies with bar size and slab thickness, but the Federal Highway Administration (1990) says it should never be less than 100 mm (4 in) or $2\frac{1}{2}$ times the biggest aggregate, or more than 225 mm (9 in). Some floor designers break that rule at the upper limit, however. They specify bigger bars to allow a spacing of about 375 mm (15 in) or so, which creates a gap big enough for the workers to step through and stand on the sub-base.

Opinions vary on the ideal depth of reinforcement. In roadbuilding, steel is usually placed near mid-depth. The Federal Highway Administration (1990) recommends that the distance from floor surface to steel should be between 1/3 and 1/2 the slab thickness, and at least 64 mm ($2\frac{1}{2}$ in). Some floor designers stick to that recommendation, but more put the steel higher in the slab to reduce crack width at the floor surface. Interior floor slabs need less cover (distance from surface to steel) than do road pavements, because floors are not exposed to the weather. The typical design has reinforcing steel 50 mm (2 in) below the floor surface. In one-way floors, the longitudinal bars normally go on top of the transverse ones.

Continuity matters. If the reinforcement is not effectively continuous for the full length of the floor, a huge crack can develop at the point of discontinuity. Bar ends should be lapped at least 25 bar diameters. The Federal Highway Administration (1990) recommends that the laps should be either staggered or skewed. In the staggered pattern, no more than a third of the bars can end in the same transverse plane, and groups of bar ends must be separated longitudinally by at least 1.2 m (4 ft). In the skewed pattern, the bar ends must lie along a line angled at least 30° from the transverse.

Where welded mats replace individual bars, staggered or skewed laps become impractical. Effective laps are still possible, but require the utmost care. Failures have occurred where the laps were inadequate, and also where adjacent mats were well lapped but separated vertically, allowing an uncontrolled crack to make its way through the gap between mats.

Even a so-called jointless floor usually contains a few construction joints. Some designers run the reinforcement through the construction joints. That keeps the joint tight but complicates the formwork. Other designers stop the reinforcement and use smooth dowels. That simplifies construction, but can result in unusually wide joints that continue to open for a long time, much like the joint at the end of a post-tensioned slab.

Sub-slab friction

Reinforced jointless floors do not rely on low friction between concrete and sub-base. Some people believe high friction is good because it encourages a closer crack spacing, with correspondingly narrower cracks.

These floors do not need slipsheets. However, if conditions call for a damp-proof membrane, there seem to be no fatal drawbacks to putting it directly under the slab, even if that results in a low coefficient of friction.

Benefits and drawbacks

Reinforced floors without joints have much to offer, especially in heavily-trafficked warehouses and factories. Benefits include:
- few or no joints to fill or seal;
- little curling;
- low maintenance costs;
- high load capacity, since almost all loading occurs in the internal position, away from slab edges and corners.

Drawbacks include:
- widespread cracking, which some users object to even though reinforcement keeps the cracks tight;
- the cost of the reinforcing steel and the labour to install it;
- difficulty in driving concrete trucks on site;
- difficulty in using laser screeds.

To get around the last two drawbacks, involving the use of concrete trucks and laser screeds, some floorlayers leave the reinforcement on the ground till the last possible minute. They drive trucks and laser screeds over the steel. Then, just before the concrete hits the ground, they raise the reinforcement and support it on chairs. This can work well, but it requires careful supervision to make sure the steel ends up at the specified height.

Steel fibres

Steel fibres can replace conventional bar reinforcement in a jointless floor. But we cannot look to the highway people for guidance here, since steel fibres see little use in roadbuilding.

Most steel-fibre jointless floors, at least in the UK and America, are designed in collaboration with the fibre manufacturers, who – no surprise – do not all recommend the same thing. Few guidelines are available for the designer who wishes to go it alone, but we can get clues by observing what's been done.

The main decisions involve fibre dosage and fibre length.

Dosages exceed that commonly used in jointed floors. Rates start at about 40 kg/m^3 (70 lb/yd^3) and top out around 60 kg/m^3 (100 lb/yd^3). Fibre length is controversial. One school argues for the longest possible fibres – 60 mm (2-3/8 in) is the practical maximum – on the grounds that they work better to keep cracks from opening wide. A rival school prefers shorter fibres – 25 mm (1 in) – because they result in very high fibre counts (number of fibres per unit volume of concrete) that prevent visible cracks by stopping micro-cracks before they penetrate the slab. They can't both be right, but both approaches have produced successful floors.

Reinforced floors with joints

These rely mainly on joints to control cracks, but also include steel reinforcement. The reinforcement serves these purposes:

- to limit the width of any cracks that occur between the joints;
- to improve load transfer at any cracks that occur between the joints;
- to improve load transfer at joints (but only if it goes through the joints, and in some designs it does not).

Most floors nowadays fall in this category, but some of them might be more profitably designed along other lines. If a floor will get little or no traffic, why not take out the reinforcement and rely on joints alone?

At the other extreme, if a floor must withstand heavy industrial traffic, why not take out the joints (the source of so many floor complaints) and rely on heavy reinforcement? But between those extremes, reinforced floors with joints have their place.

Nominal reinforcement

Some designers lay out joints in a pattern that would be suitable for an unreinforced floor, and then add very light reinforcement as insurance against cracks. This is nominal reinforcement – reinforcement in name only. The steel usually takes the form of light wire mesh. Some UK builders have used A49 mesh, which provides 49 mm^2/m (0.023 in^2/ft) of steel in two directions. In a 150-mm (6-in) slab, that amounts to only 0.03% steel in each direction. Even lighter mesh is used in America.

It seems unlikely that such light reinforcement has much effect beyond reassuring those people – and they are many – who believe a concrete floor simply must contain some reinforcing steel.

Light reinforcement for joint stability

In this method, light reinforcement goes through the induced joints (usually stopping at construction joints). The reinforcement reduces joint opening and improves load transfer at joints and cracks.

The joint spacing is similar to that used in totally unreinforced floors (see Table 16.1), though some designers stretch it a little.

The amount of reinforcing steel is critical, though opinions vary on how much is needed. Too little, and the floor might as well be unreinforced. Too much, and the joints cannot relieve enough stress to prevent cracks. The usual amounts range from about 0.1% to 0.2%. In America, many designers have standardized on 12-mm bars spaced on 450-mm centres (#4s at 18 in), each way. That pattern gives 0.18% steel in a 150-mm (6-in) slab, and 0.14% where slab thickness is 200 mm (8 in).

The reinforcement normally goes in the top half of the slab, about 50 mm (2 in) from the surface. It can take the form of either individual bars or wire mesh. Bars are easier to keep at the desired height, especially if the spacing lets workers step between the bars and stand on the ground (Figure 16.3).

Figure 16.3 Light reinforcement for crack control in a jointed slab – 12-mm (#4) bars spaced 600 mm (24 in) on centre

Reinforcement to allow greater joint spacing

A common use of reinforcement is to allow a wider joint spacing than would be safe in an unreinforced floor. The joint spacings are typically 6–8 m (20–25 ft), but may range up to 12 m (40 ft).

What we need to remember here – and what many people never learn – is that the reinforcement does not prevent cracks. Its job is to limit the width of cracks after they have occurred.

Some designers use the following equation, called the drag factor formula, to determine how much reinforcing steel to use:
In SI units:

$$Q_s = \frac{(1.57FL)}{f_y}$$

where:

Q_s = required steel as a percentage of the slab's cross-sectional area
F = coefficient of friction between slab and base

L = slab length, in m
f_y = characteristic strength of steel in MPa

In fps units:

$$Q_s = \frac{(69.5FL)}{f_y}$$

where:

Q_s = required steel as a percentage of the slab's cross-sectional area
F = coefficient of friction between slab and base
L = slab length, in ft
f_y = characteristic strength of steel in psi

The coefficient of friction, F, is often assumed to be 1.5. But the real value varies (see Table 16.2). Putting a polyethylene slipsheet under the slab can reduce the coefficient to 0.5.

The length of slab, L, is either the total length of the floor or the distance between free contraction joints, whichever is less. Unless the floor is square, we need to look at L in both the longitudinal and transverse directions.

Typical reinforcing steel has a characteristic strength of 460 MPa in the UK, and 60 000 psi in America.

Bars versus mesh

The reinforcement can consist of individual deformed bars or wire mesh. Bars offer more flexibility, since the designer can choose whatever spacing gives the desired percentage of steel. Mesh saves labour. British designers use mesh almost exclusively, and can choose between A-type with the same amount of steel in both directions, and B-type with heavier reinforcement in the longitudinal direction. B mesh finds use in narrow-strip pours (less common now than in past decades) which need more steel in the

Table 16.2 Sub-slab friction (Post-Tensioning Institute)

Material Directly Under Slab	Coefficient of Friction
Polyethylene sheet	0.5–0.9
Sand	0.7–1.0
Granular sub-base	0.9–1.7
Plastic soil	1.3–2.1

longitudinal direction. American designers are split between bars and mesh, with some preferring bars on the grounds they are more easily kept in place during the pour.

If using individual bars, you can simplify construction by choosing big enough bars so their spacing exceeds 350 mm (14 in). This is technically inferior, because many small bars reinforce more efficiently than a few big ones. But it's worth doing, because the effect on constructability outweighs the small reduction in reinforcing efficiency.

If using mesh, take care to ensure it ends up at the specified height within the slab. That can be hard to do, especially with light-gauge material that yields under foot traffic. Some floorlayers like to leave it on the ground till it is covered with concrete, and then pull it up with hooks – but with that method the mesh's final position is anybody's guess. It's far better to support the mesh on chairs or bolsters.

Continuity of reinforcement

Reinforcement can continue through the joints, or not.

If it goes through, then the joints become tied. They still act as hinges to relieve curling stresses, but they do little to relieve straight shrinkage stresses (unless the steel yields). When figuring the slab length (L) for the drag factor formula, you ignore the tied joints and measure all the way to the slab end or the first joint that is not tied. Running the reinforcement through the joints improves joint stability and reduces joint maintenance, but increases the risk of cracks between the joints.

If reinforcement stops at the joints, then they become free contraction joints. They may need dowels for load transfer, depending on loads and traffic patterns. Stopping the reinforcement at the joints reduces the risk of cracks, but increases the likelihood of joint problems.

Steel fibres

In recent years steel fibres have increasingly replaced steel bars and wire mesh in jointed floors. Depending on the dosage rate, fibres can serve as nominal reinforcement, as reinforcement to stabilize joints, or as reinforcement to allow greater joint spacing.

At doses under 15 kg/m^3 (25 lb/yd^3), steel fibres act as nominal reinforcement. They do not cost much, and they do not do much. Such reinforcement can replace light wire mesh – but in many cases a more reasonable change would be just to take out the mesh and replace it with nothing.

At higher doses, fibres help stabilize joints and allow wider joint spacings. Guidelines are hard to find, however, and the drag factor formula is of little use with fibres. (Some people do not think it's all that useful with conventional reinforcement, either.) Most designers rely on the steel-fibre suppliers for help.

Prestress

This method puts the concrete slab in compression before drying shrinkage and thermal contraction can create the tensile stresses that cause cracks. If the compressive prestress equals or exceeds the tensile stress induced later, the floor will not crack. Actually, the prestress could even be slightly less than the tensile stress, since concrete has some ability to withstand tension before cracking.

Two very different prestressing methods are widely used in floors: shrinkage-compensating concrete and post-tensioning with steel tendons. Pre-tensioning with steel tendons – the usual method for prestressing precast elements, and the process that normally comes to mind when someone mentions prestressed concrete – is rarely if ever used for ground-supported floors.

Shrinkage-compensating concrete

This is made with expansive cement, or with ordinary cement combined with an expansive addition. The concrete expands slightly soon after setting. After that early expansion, it undergoes drying shrinkage at a normal rate.

If we restrain the expansion, it creates compressive stresses within the floor to offset the tensile stresses that come later. The usual way to provide that restraint is with reinforcing steel, though sub-slab friction plays a contributing role.

Shrinkage-compensating concrete is rare in the UK, and British standards say almost nothing about it. It is more common in America, and so we need to look to American standards. ACI 223 provides general guidelines for the design and use of the material. ASTM C 845 specifies expansive cement, while ASTM C 157 and ASTM C 878 describes standard tests for measuring the expansion of concrete specimens.

The main key to success with shrinkage-compensating concrete is to make sure the expansion exceeds the shrinkage. Floorlayers who specialize in this field normally test a trial batch for expansion. That is an important step because the amount of expansion depends not only on the expansive cement, but also on the properties of the other components, which vary from joint to job. Testing the trial batch for drying shrinkage provides added safety. If a mix shows unusually high drying shrinkage, the normal amount of expansive cement may not suffice to prevent cracks. The solution is to adjust the mix so it expands more or shrinks less, or both.

The standard rules call for shrinkage-compensating concrete to be laid out in slabs that are square, or nearly square. Each slab, which can be as big as 30 × 30 m (100 × 100 ft) is placed by the large-bay method and contains no joints, except isolation joints at building columns and other penetrations. A few engineers have questioned the requirement for square slabs. If they are right, it would open the possibility of narrow-strip floors, including superflat floors, made of shrinkage-compensating concrete. But for now square panels remain the norm.

Because the prestress from expansive cement is small, it is not considered to add directly to the floor's structural strength.

The main drawback to shrinkage-compensating concrete is its price. If expansive cement cost the same as ordinary Portland cement, it might well have become the standard material. But expansive cement always costs more.

Shrinkage-compensating concrete finds its greatest use in busy warehouses, particularly in the grocery trade. The typical owner choosing shrinkage-compensating concrete has concluded that the savings in joint maintenance, which last for the life of the warehouse, offset the initial cost.

Post-tensioning

With this method, the prestress comes from steel cables called tendons. The "post" in post-tensioning refers to the fact that the tendons are stretched after the concrete has set, in contrast with pre-tensioning in which the tendons are stretched before concrete is placed.

Post-tensioning makes possible the construction of very big slabs with no joints, and with few or no cracks. The longest practical dimension is about 130 m (400 ft). Unlike slabs made of shrinkage-compensating concrete, post-tensioned slabs are not strictly limited as to shape. They can be long and narrow, or L-shaped, with no compromise in performance.

Bonded and unbonded tendons

Tendons can be bonded (also called grouted) or unbonded. In bonded construction, you install each tendon within a rigid tube. After the tendon has been fully stretched, you pump grout into the tube, locking the tendon in place along its whole length. In unbonded construction, each tendon comes greased and encased in a plastic sheath. The tendon sticks out beyond the sheath at each end so it can be stretched and anchored. An unbonded tendon applies stress to the concrete only through the anchors at each end, like a giant clamp.

Bonded construction offers flexibility in floor usage, since the tendons can be cut later without much effect. Unbonded construction costs less and resists corrosion better. Regional preferences vary, with bonded construction the norm in Australia and parts of Asia, but unbonded carrying the day in America.

Tendon details

In ground-supported floors, tendons normally run straight and lie near the slab's mid-depth. This differs from normal suspended- floor practice in which tendons are draped – curved in the vertical plane – to resist bending moments. Tendons can run in one or two directions. If they run in one direction only, the floor needs some other method – usually closely spaced joints – to control cracks in the other direction. One-way post-tensioning is used in racked warehouses where the traffic aisles all run the same way (Figure 16.4).

The typical tendon consists of seven steel wires – one down the centre, and six more spiralling around it. The steel is much stronger than that used for ordinary reinforcement; in America the standard product is Grade 270 steel, with characteristic strength of 1830 MPa (270 000 psi). Table 16.3 shows key properties for three tendon sizes common in America. Post-tensioning steel costs more than reinforcing steel, but you usually need a lot less of it to accomplish the same goal.

At each end the tendon passes through a steel anchor. One type, called a barrel anchor, mounts outside the slab and is installed after the edge form has been stripped. The more common internal type mounts just inside the slab's edge, and is nailed to the edge form. Wedges connect the tendon to the anchor, preventing movement toward the centre of the slab, but allowing movement in the other direction so the tendon can be stretched. Most designs include backer bars – two 12-mm (#4) deformed bars lying parallel to the tendons and close to the anchors (see Figure 16.5).

Figure 16.4 One-way post-tensioned floor with unbonded tendons

Table 16.3 Properties of steel tendons for post-tensioning (Lang Tendons, Inc.)

Property	Strength Designation		
	23K	41K	58K
Nominal diameter	10 mm	13 mm	15.2 mm
	3/8 in	1/2 in	0.6 in
Steel area	54.8 mm²	101.3 mm²	140.0 mm²
	0.085 in²	0.157 in²	0.217 in²
Ultimate tensile strength	102 kN	184 kN	261 kN
	23.0 kips	41.3 kips	58.6 kips
Maximum jacking force	82 kN	147 kN	209 kN
	18.4 kips	33.0 kips	46.9 kips
Design load	61 kN	110 kN	157 kN
	13.8 kips	24.8 kips	35.2 kips

Figure 16.5 Anchorages for post-tensioning with unbonded tendons

Stretching the tendons

Stress is applied with an hydraulic jack, which both stretches the tendon and sets the wedges. A gauge shows the applied force, so the operator can stop when the specified value has been reached. It is good practice to calculate the expected elongation and measure it during the stressing. If the measured elongation differs much from the prediction, it can be a sign of gauge error, or tendon slippage at the far end, or the tendon's hanging up between the anchors.

Slabs under 15 m (50 ft) long are usually stressed in one operation after the concrete has reached about 25 MPa (3000 psi) compressive strength. But one-step stressing doesn't work so well on longer slabs, which are likely to crack before stress is applied. Two-step stressing prevents most of those early cracks. The first step is to apply one third of the ultimate stress as soon as the concrete's compressive strength reaches 10 MPa (1200 psi), determined by testing site-cured concrete cubes or cylinders. The second step is to apply full stress at 25 MPa (3000 psi). Once in a while a crack occurs before even partial stress can be applied. If this happens, sealing the crack with sticky tape will help keep out debris and, with luck, let it close up later.

Under most conditions, early stressing takes place the day after the pour. But in hot weather it may need to occur the same day. The people testing the concrete strength, as well as the people doing the stressing, need to be ready whenever the concrete strength is expected to reach 10 MPa (1200 psi), even if that happens at night or on the weekend.

Shorter tendons can be stretched from one end; longer tendons should be stretched from both ends. One-end stressing not only saves time, but also allows the slab to be poured right up against a wall or adjoining slab, without the need to leave a gap. Aalami and Jurgens (2003) recommend two-end stressing where the length exceeds 36 m (120 ft), but they are writing about suspended floors with draped tendons. Since ground-supported floors have straight tendons, friction losses are much lower and it is probably reasonable to stress from one end at much greater lengths.

Calculating the prestress

Because post-tensioning applies force at the slab edges, the prestress is always greatest there. It gradually diminishes with distance from the anchors, reaching its minimum at the slab's mid-length. Designers proportion the tendons (size and spacing) to provide a certain residual prestress at the mid-length, after allowance for friction losses. Where the post-tensioning serves only to control cracks, the minimum prestress ranges from 350 kPa (50 psi) to 1.0 MPa (150 psi). The Post-Tensioning Institute (1980) suggests 350 kPa (50 psi), but that leaves little room for error.

Prestress above 1.0 MPa (150 psi) is specified where the goal includes an increase in the floor's load-bearing capacity. If you design a floor for mid-length prestress of, say 2.0 MPa (290 psi), that is effectively the same raising the concrete's flexural strength by that amount. You can take advantage of that to reduce slab thickness or use lower-strength concrete, compared to a non post-tensioned slab. In warehouse construction, post-tensioning can often save 50 or 75 mm (2 or 3 in) on slab thickness. Sometimes this pays for the post-tensioning, providing excellent crack control at no net cost. But the savings are not available on lightly loaded floors, which are usually designed to minimum thickness requirements anyway and cannot be made thinner.

The following equation determines the tendon spacing:

$$S = \frac{P}{[(C_{mid}A) + (W_{slab}0.5LF)]}$$

where:

S = tendon spacing in m (or ft)
P = design tensile stress on each tendon, in N (or lb)
C_{mid} = desired mid-length compressive stress in MPa (or psi)
A_{slab} = cross-sectional area of slab per 1 m (or ft) in mm² (or in²)
W_{slab} = weight of slab per m² (or ft²) in N (or lb)
L = slab length between anchors in m (or ft)
F = coefficient of friction between slab and ground

Workmanship

More than any other method of crack control, post-tensioning requires careful workmanship. Success hinges on attention to three details: isolation joints, sub-base preparation and concrete compaction.

A post-tensioned floor must be well isolated from other building elements so it can shrink freely. Isolation joints must be wider than usual because slab movement is large, especially near the edges. A big post-tensioned slab can shrink by 30 mm (over 1 in) at each end, and the isolation joints must be capable of accommodating that. It's good practice to isolate the slab with soft foamed plastic at least twice as thick as the predicted movement.

The sub-base should be smooth and even to reduce friction. It is particularly important to eliminate any sharp steps or ruts, because they tend to pin the slab and keep it from sliding.

Last, concrete must be well compacted around the anchors. Compaction matters in every concrete floor, but post-tensioning raises the stakes since poor compaction can result in a highly-stressed anchorage breaking out.

Benefits and drawbacks

Post-tensioned floors offer these benefits:

- very good crack control, with most slabs remaining free of visible cracks;
- increased load-bearing capacity (depending on the amount of prestress);
- few restrictions on slab shape or size.

They also have drawbacks that limit their use:

- lack of flexibility in floor usage (especially with unbonded tendons, which should never be disturbed after they have been stretched);

- need for gaps at slab ends, to leave access for the stressing jack;
- very wide joints where the post-tensioned slab ends, since all the concrete's shrinkage is concentrated there.

Sub-slab friction

An important factor in crack control is the friction between the concrete slab and the layer right below it. Reinforced floors without joints do not depend on low friction, and may even perform better if the friction is high. But with that exception, low friction is desirable and offers these benefits:

- Unreinforced floors are less likely to develop random cracks.
- Reinforced floors with joints may need less reinforcement.
- Post-tensioned floors need fewer tendons.

Table 16.2 shows friction coefficients for materials often used under slabs. A polyethylene slipsheet provides the lowest friction.

There are drawbacks to polyethylene, however. By preventing moisture loss at the bottom of the slab, it promotes bleeding and curling, and can lengthen finishing time. To reduce those problems, some builders punch holes or cut slits in the polyethylene sheet – provided the polyethylene is not expected to do double duty as a damp-proof membrane. The holes or slits let water pass but have little effect on slipperiness.

To reduce friction further, some designers call for two layers of polyethylene, sometimes with water between them. Such double slipsheets are common under post-tensioned floors, where it is usual, if somewhat optimistic, to assume a friction coefficient of 0.4.

Part V

The Floor Surface

17

Floor Finishing

One of the biggest differences between floors and other building elements made of concrete is that floors must be finished. With walls and columns, the surfaces exposed to view are cast against forms. A floor also has forms, but its all-important top surface is unformed. Finishing is the process of making that unformed top into a usable floor surface.

Floor finishing is much more than a simple smoothing process. Though little appreciated, it is one of the hardest, most highly-skilled jobs in the building trades.

This chapter covers:

- principles of finishing;
- tools;
- sequence of finishing steps;
- types of finishes;
- other finishing methods;
- how to specify finishes.

Principles of finishing

Ordinary concrete finishing consists of three basic steps: striking off, floating and trowelling.

Striking off, also called screeding, is the process of levelling the fresh concrete. It takes place right after the placing and compacting operations described in Chapter 11. Accuracy in strike-off is the key factor in making floors level.

Floating is the process of working the concrete with wide-bladed tools held flat against the floor surface. Floating serves these three purposes:

- It depresses the coarse aggregate to leave a surface mortar of cement, fine aggregate and water.
- It smooths the small ridges and holes left after striking off.
- It compacts the surface mortar.

On some floors the finishing ends with floating. This leaves a rough surface called a float finish, with the texture of coarse sandpaper.

Trowelling further smooths and compacts the surface mortar, using steel blades held at an angle to the floor surface. Early trowellings are done with blades held almost flat. Later trowellings are done with blades at progressively steeper angles, increasing pressure on the concrete. Repeated trowel passes produce a sort of case-hardening, making the floor surface hard and long-wearing.

Tools for finishing

Catalogs show dozens of tools for finishing concrete floors. The choice is, to an extent, arbitrary and personal. One floorlayer likes a wooden hand float; another swears by magnesium. Nevertheless, there are a few basic types that most floorlayers use. We can divide tools into these groups:

- tools for striking off;
- floats;
- trowels;
- other tools.

Tools for striking off

These include hand straightedges, highway straightedges, vibrating screeds, roller screeds and laser screeds.

Hand straightedge

Also called saw rod or saw beam. This is the most basic tool in floor construction. It was undoubtedly used on the earliest concrete floors, and

Floor Finishing

yet it plays a key role in modern superflat construction. It is nothing more than a straight (some straighter than others) beam, usually rectangular in section. Typical dimensions are 25–50 mm (1–2 in) wide, 90–150 mm ($3\frac{1}{2}$ – 6 in) deep and up to 6 m (20 ft) long. Some straightedges have short handles at each end; most do not.

Timber beams are common, but extruded metal tubing is straighter and lasts longer. There is a minor controversy over the use of aluminium versus magnesium straightedges. Both metals are light and lend themselves to being extruded into precise shapes, but aluminium particles can cause a damaging reaction in fresh concrete, raising blisters on the floor surface. It is far from certain, however, that enough aluminium ever rubs off a straightedge to cause this reaction.

Hand straightedges are common in narrow-strip construction, where they easily span the pour and rest on the side forms. They are also used for wet screeding on large-bay pours (see Figure 17.1).

Highway straightedge

This is a special type of hand straightedge with a long handle attached to its midpoint. The name comes from its origin in road paving. Highway straightedges make floors flatter.

Figure 17.1 A hand straightedge is used for wet screeding

Blade lengths range from 3 to 5 m (10 to 16 ft). Blade cross-sections range from 40 by 40 mm (by $1\frac{1}{2}$ by $\frac{1}{2}$ in) to 50 by 125 mm (2 by 5 in). Lighter highway straightedges are sometimes called check rods, because they serve to check a floor's flatness but cannot move much material. Heavier straightedges are sometimes called bump cutters or scraping straightedges.

The connection between handle and blade is adjustable. With the better models, you can adjust the blade's angle by twisting the handle.

The highway straightedge strikes off at right angles (or sometimes diagonally) to the primary strike-off. An essential tool in narrow-strip super-flat work (see Figure 17.2), it is also used on wide laser-screeded pours to smooth out the seams between adjacent passes of the laser screed's auger.

Vibrating screed

This strikes off the concrete and also provides some compaction. The three main types are single-beam, double-beam and truss.

A single-beam screed is just a beam – timber or metal tubing – with a vibrator bolted on. The beam is typically 50 by 150 mm (2 by 6 in) in section.

Figure 17.2 Highway straightedge

Figure 17.3 Double-beam vibrating screed

A double-beam screed is similar but includes two beams, one in front of the other (see Figure 17.3). Single-beam and double-beam screeds are limited in length to about 6 m (20 ft). At that length their use is largely confined to narrow-strip pours. Vibration comes from an electric motor or small internal-combustion engine. Beam vibrators are usually pulled along by hand.

A truss screed is made of steel and looks like a scale model of a truss bridge (see Figure 17.4). It comes in sections that can be bolted together to span wide pours. Some truss screeds span as far as 25 m (80 ft) and can be adjusted for camber. However, spans of 5 to 12 m (16 to 40 ft) are more common. Power comes from an internal-combustion engine or, less often, compressed air. Vibration is applied at several points along the screed's length. Winches – either hand-cranked or powered the screed's engine – pull the screed along.

All vibrating screeds are used in much the same way. They rest on the side forms, or on intermediate screed rails, and are pulled along slowly. It is important to get the speed right. If the screed goes too fast it leaves high and low spots. If it goes too slow it may overvibrate the concrete, bringing too much watery mortar to the surface.

Figure 17.4 Truss-type vibrating screed

Roller screed

This is a steel tube that spins at high speed, rotating against the direction of travel. It is pulled along the tops of the side forms, flicking forward any concrete that lies above the finished floor level. Roller screeds come in lengths up to about 9 m (30 ft), making them suitable for narrow-strip pours and some wide strips. A drawback is the fixed length; unless a contractor is prepared to buy more than one roller, all slabs need to be cast at about the same width.

Roller screeds work on very stiff mixes, making them a reasonable choice for steeply sloped floors. Such floors are not common, however.

Unlike vibrating screeds and laser screeds, roller screeds do not vibrate. If you strike off with a roller screed, count on using another method, probably poker vibrators, to compact the concrete.

Laser screed

This levels the concrete with a vibrating auger that is adjusted automatically relative to a laser beam. In large-bay construction, the laser screed controls levelness far better than any other strike-off tool (Figure 17.5).

Since its introduction in the 1980s, the laser screed has transformed the laying of big industrial floors. Contractors with laser screeds achieve rates of production and standards of flatness and levelness that would have seemed absurd a generation ago. With laser screeding, it is now common to lay as much as $3000\,m^2$ ($30\,000\,ft^2$) a day, and to achieve overall F-numbers of $F_F 50/F_L 30$ or even $F_F 60/F_L 40$.

Figure 17.5 Laser screed

Because the laser screed does not rest on the side forms, it can be used on slabs of almost any width. Quality suffers on the widest pours, however, because of the amount of time it takes to work the screed all the way across. For good flatness in hot-weather pours, it is wise to limit pour width to about 25 m (80 ft). In cool weather, wider pours are safe.

Despite its powerful advantages, the laser screed faces some limitations. It works best on ground-level floors. The standard laser screed is too heavy for most suspended slabs, and the lighter, smaller laser screeds developed for upstairs work are both less productive and less precise.

Another limitation is the difficulty in working the laser screed around reinforcing bars and post-tensioning tendons. If the bars or tendons are chaired up before the pour, there is no place for the laser screed to stand. The usual solution is to leave the steel on the ground till the last possible minute, but that complicates the pour and creates a risk that the bars or tendons may not end up in the right position. Another solution is to support the screed on a moveable bridge. Laser-screed operators often prefer pours with nothing in the way; this is a key but not always recognized factor in the current popularity of unreinforced and steel-fibre-reinforced slabs.

A third limitation is the screed's inability to level a floor all the way to the edge. With every other method of striking off, when you get to the slab's edge the tool rides on the side form. A laser screed cannot do that because its spinning auger would grind against the form. Laser screeding normally stops about 0.5 m (1–2 ft) from the edge, leaving a narrow strip to be worked by hand. The result can be a floor that is very level over most of its area (where the laser screed did its job), but with severe slopes or bumps at the construction joints.

Floats

A float has a flat surface that is held parallel to the floor surface. Edges are rounded so they do not dig in. Floats can be divided into two groups according to when they are used. The first group consists of bull floats and darbies that are used right after striking off, before bleed water appears. The second group includes hand floats and power floats used later, after all bleed water has disappeared.

The *bull float*, also called skip float, is a wide tool with a long handle. The typical model is about 200 mm (8 in) wide and 1–1.5 m (3–5 ft) long. Nowadays most are made of aluminium of magnesium, but some floorlayers prefer the traditional wood. The handle consists of tubular sections that snap or screw together. By adding sections you can make the handle as long as you like, but it is hard to maneuvre a handle over 6 m (20 ft) long. A long handle lets the floorlayer smooth the floor without walking in the fresh concrete. At the time bull floating normally occurs – right after striking off – the concrete is far too soft to support a person's weight.

A newer but related tool is the *channel radius float*. Longer and stiffer than a bull float, it is somewhat harder to use but does a better job making floors flat.

A *darby* is a long float with a low handle. Typical dimensions range from 50 to 100 mm (2 to 4 in) wide and from 1 to 1.5 m (3 to 5 ft) long. It can be made of wood or magnesium. The handle is usually located at one end so the floorlayer can sweep the tool across the floor while kneeling at the slab's edge. Some darbies have two handles, one at each end. Darbies can replace bull floats on very narrow slabs. Some floorlayers use darbies at the slab's edges, and bull floats farther out.

The *highway straightedge*, while not usually considered a floating tool, often replaces bull floats and darbies. See the preceding section on tools for striking off.

The basic *hand float* resembles a darby, but is shorter and is used later in the sequence of finishing steps. Sizes vary greatly, but the typical float is about 90 mm ($3^1/_2$ wide) and 300–600 mm (12–24 in) long. Common materials are wood (traditional), magnesium and resin-impregnated canvas. It is possible to float a whole floor by hand, but nowadays this only occurs on the smallest jobs. Bigger floors are usually power floated, but even so hand floats are used at edges and in congested areas.

Power floats come in two types. The original type, not seen much these days, is designed strictly for floating, not trowelling. It consists of a steel disc about 460 mm (18 in) in diameter drive by an internal-combustion engine. This type incorporates vibration as well as rotary action. It is heavy and cannot be used till the concrete is quite hard.

Today the tool of choice is a power trowel machine (described below) fitted with attachments designed for floating. Any trowel machine can be turned into a power float by attaching to it either float shoes, combination blades or pans.

Float shoes are individual steel floats that fit over the trowel blades. Each float shoe is turned up on both the leading and trailing edges. Combination blades are turned up on the leading edges only. They can be used for floating (if held flat against that floor) and for trowelling (if tipped). Combination blades represent a compromise – not quite so good at floating as float shoes or pans, and not quite so good at trowelling as dedicated trowel blades.

A pan is a steel disc with a turned-up edge all around. Pans can be fitted to almost any trowel machine, but it's not a good idea to put them on walk-behind machines with a diameter greater than 1 m (3 ft), because friction makes the bigger machines hard to control. (The big walk-behind machines work fine with float shoes, however.) Pans are always dished, with the edge 3–10 mm (1/8–3/8 in) higher than the centre. At least one manufacturer offers pans with minimal dish, for use on floors that must be very flat.

The 1990s saw a big increase in the use of riding trowel machines with float pans. The machines have grown bigger, with some new running pans of 1.5 m (6 ft) diameter. The bigger pans have a good effect on floor flatness.

Trowels

These are tools with flat surfaces held tipped to the floor surface. Trowels resemble floats but there are some clear differences. Trowels have sharp edges; floats have rounded edges. Trowels are used at an angle to the floor surface; floats are held level. Trowels are made of steel; floats were traditionally made of wood, though other materials are used today.

Three kinds of trowels are in wide use: hand trowels, fresnos and power trowel machines.

Hand trowels

A hand trowel is a thin sheet of steel, usually spring steel or stainless, with a low handle. It is typically 75–125 mm (3–5 in) wide and 250–500 mm (10–20 in) long. Trowels for floors are almost always rectangular; pointed bricklayer's trowels have no place in floor construction. Trowel size is largely a matter of personal preference, but in general it is good practice to use the biggest practical size for the first trowel pass. Smaller trowels may be used for later passes as the concrete stiffens. Some floorlayers carry a very small hand trowel for filling in holes.

Though it is possible to produce a smooth, hard surface with hand trowels, power trowel machines do a better job. In modern industrial floor construction, most floorlayers rely heavily on trowel machines and floors are the better for it.

Fresnos

A fresno is a large trowel with a long handle. It can be thought of as the trowelling equivalent of a bull float. A typical fresno is 125 mm (5 in) wide and 600 mm–1m (24–40 in) long. It works fairly well for a first trowelling on soft concrete, but is not very effective later as the concrete hardens. For that reason, fresnos are seldom used where very smooth finishes are needed.

Power trowel machines

These are rotary devices that resemble ceiling fans. Each rotor turns several (usually four) trowel blades. Power normally comes from an internal-combustion engine, but electric machines are available for use where exhaust fumes are not acceptable. You can turn any power trowel into a power float by fitting it with float shoes, combination blades or pans.

Walk-behind trowel machines have a single rotor and a long handle sticking out to one side (see Figures 17.6 and 17.7). Sizes vary, but the most-used

Figure 17.6 Walk-behind power trowel machine

288 Design and Construction of Concrete Floors

Figure 17.7 This small trowel machine is used at the slab's edge

machines come in diameters of 910 and 1170 mm (36 and 46 in). Smaller machines are used at edges and in tight quarters.

Riding trowel machines have two or (rarely these days) three rotors. Rotor diameters range up to 1.5 m (5 ft). Dual-rotor riders with the biggest blades are sometimes called "10-foot machines" because they sweep a 10-foot (3-m) swathe with each pass (Figure 17.8).

Higher productivity is the main reason for using riders, but not the only one. Because they weigh more than walk-behind machines, they exert more pressure on the blades. This extra pressure allows some finishing steps to take place later than would otherwise be possible, and helps achieve a

Figure 17.8 Riding trowel machine fitted with float pans

dense, burnished finish. On the other hand, the weight and power of a big riding trowel machine sometimes encourages sloppy timing.

Any trowel machine worth buying provides for changing the blade angle and speed of rotation. For floating, blades stay level and spin slowly. For the first trowelling, they are tilted slightly but do not speed up much. Blade angle is then increased for each later trowel pass. Speed is also increased as the concrete become harder.

Power trowelling has become the standard method for finishing industrial floors.

Other tools

Floorlayers use a few finishing tools that do not fit neatly into the three categories of striking off, floating and trowelling.

A *grate tamper* is a metal sheet full of holes, mounted on a rectangular frame, with a tall handle. The floorlayer pushes the grate down into the concrete to depress the coarse aggregate. Because grate tampers have a bad effect on surface regularity, they should never be used on floors that need a high or even moderate degree of flatness or levelness. It's questionable whether they are needed at all in modern floor construction.

A *finishing broom* is used where a broom finish is specified. This is more common on outdoor paving than on floor slabs. The broom is 600–120 mm (24–48 in) wide and has soft, fine bristles. It needs a very long handle so a floorlayer can drag it over the floor with walking in the concrete.

A *wire broom* has stiffer, coarser bristles. It is used to produce a scratch finish, which is sometimes specified for slabs that will get toppings or ceramic tile.

Grinders are used both to correct defects and as a finishing tool in their own right.

Some builders try to repair floors with hand-held angle grinders, but such machines are useless for all but the smallest defects. A better choice for correcting curled-up joints and rough spots is a concrete grinder with a 250-mm (10-in) diamond disc. This machine is about the size of an ordinary lawn mower, and it typically powered by a 6-kW (8-hp) internal-combustion engine. Electric models are also available. Most diamond-disc grinders include a hose fitting so water can be introduced at the grinding face, but some have vacuum systems to remove dust. Bigger diamond-disc grinders are sometimes used to repair large areas.

Early-age grinding (described below) uses a different sort of machine. The standard tool is a grinder with two rotating discs, each fitted with three abrasive blocks called stones. Stone grinders have little effect on hard, strong concrete. They work in early age grinding because that process calls for their use while the concrete is still young and weak.

Special high-speed grinders are used to finish terrazzo.

A *jointing tool* is a hand tool for inducing joints. It is also called a groover, a name that describes its use perfectly. It is a small hand float, typically about 150 mm (6 in) long, with a ridge on the bottom. This tool has little application in modern floor construction; there are better ways to make joints.

An *edger* is a hand tool for rounding off the edge of a slab. It resembles a jointing tool (the two often come in matched pairs) but with the ridge on one side instead of the centre.

The order of finishing steps

The finishing of a concrete floor takes many separate steps. The exact number and order of steps depend on several factors, including the type of finish desired, the workability and finishability of the concrete and the skill of the floorlayers.

Here is a typical sequence:

1. Strike off with a vibrating screed.
2. Use a 3-m (10-ft) highway straightedge at right angles to the screed. Lap each pass by 1.5 m (5 ft).
3. Wait for bleed water to appear and evaporate. This can take several hours.
4. Even after bleed water had evaporated, wait for the concrete to stiffen enough so that a footprint goes no more than 5 mm ($^1/_4$ in) deep.
5. Float, using pans on a power trowel machine.
6. Use a hand float at edges and penetrations.
7. Wait till the concrete has hardened enough for trowelling. Experience is the only reliable guide here.
8. Power-trowel, using trowel blades held almost level.
9. Use a hand trowel at edges and penetrations.
10. Power-trowel a second time, with blades tilted.
11. Use a hand trowel again at edges and penetrations.
12. Power-trowel a third time, with blades tilted higher and throttle partly opened.
13. Use a hand trowel again at edges and penetrations.
14. Power trowel a fourth time, with blades steeply tilted and the throttle wide open.
15. Start curing right away.

The sequence described here would be suitable for a direct-finished industrial, laid in wide or narrow strips and designed to meet these surface requirements:

- surface regularity – $F_F 30/F_L 20$ overall (ACI 117 "flat" category);
- wear resistance – AR2;
- description of surface – burnished finish;
- slip resistance – not critical.

Bear in mind that any list of steps is descriptive, not prescriptive. Some floorlaying crews achieve excellent results using methods different to those in my list.

Types of finishes

Floor finishes are often described in terms of the last finishing tool used on them. We shall consider these types, from smoothest to roughest:

- smooth trowel finish;
- float finish;

- broom finish;
- scratch finish.

Smooth trowel finish

Multiple trowellings leave a floor smooth and hard. If taken as far as it can go, trowelling produces a very dense, somewhat darkened surface; this is called a burnished or burned finish (Figure 17.9).

The sequence of finishing steps described above leads to a burnished finish.

In many ways this is the ultimate floor finish: hard-wearing, easily cleaned and good-looking. It is usually the best choice for an industrial floor.

This is not ideal for every job, however. If the floor will be covered with tile or carpet, it is a waste of time and effort to put a burnished finish on it. And if a floor will get a bonded topping or coating, a smooth finish is not only unneeded but counterproductive. The hard finish will have to be removed by shot-blasting or similar means.

Figure 17.9 Burnished finish

Even floors exposed to wear should not get a smooth trowel finish if slip resistance matters. A burnished slab can become very slippery if liquid or fine powder is spilled on it. There is seldom any problem, however, if the floor stays clean and dry.

Occasionally, a designer or user will want a finish just slightly rougher than a standard trowel finish, but smoother than the rough finishes described below. You can achieve this – with a little luck – by leaving off what would otherwise be the last trowelling operation. The surface should remain very slightly fuzzy. Designers who specify this approach should know it may not eliminate all visible trowel marks.

Float finish

In the finishing sequence described above, we could produce a float finish by dropping steps 7–14.

A float finish is rougher than a smooth trowel finish. It has better slip resistance and is most often specified for that reason. Sometimes it is chosen for its appearance; the floating operation leaves swirls, which some people find attractive.

On the debit site, a float finish has less wear resistance than a smooth trowel finish, and is harder to clean.

A float finish is usually good enough for floors that will get coverings. But if the covering will be a thin sheet material, some of the float marks might show through. In that case, it's safer to give the floor at least one trowel pass.

Broom finish

This is produced by dragging a finishing broom across the soft concrete. It is normally done after floating. In the finishing sequence described above, brooming would take place after step 6, and would replace steps 7–14.

A broom finish is functionally similar to a float finish. But it looks different because the broom leaves straight lines, in contrast with the swirl marks left by floating.

Broom finishes are seen more on outdoor pavements than on floor slabs.

Scratch finish

This is made by scoring the soft concrete with a wire broom. In the finishing sequence listed about, the scratching would replace steps 5–14.

A scratch finish is never used where it would be seen in the finished building. Its only purpose is to prepare a base slab for a bonded topping or ceramic tile.

Though successful toppings have been laid over scratch-finish slabs, there are better ways to ensure bonding. One good method is to float the slab, and later roughen the hardened concrete by shot-blasting or scabbling.

Other finishing methods

Most concrete floors are finished by the basic operations of striking off, floating and trowelling. But there are other methods used in addition to, or in place of, the basic three. Some of them are:

- early-age grinding;
- dry shakes (also called sprinkle finishes);
- vacuum dewatering;
- shot-blasting

Early-age grinding

This alternative to floating and trowelling, gets its name from the fact that the concrete is ground early on, after it has set but before it has gained much strength. The process involves these steps:

1. Strike off in the normal way.
2. Smooth with a bull float or highway straightedge.
3. Cover with polyethylene sheeting.
4. One or two days later, remove the polyethylene and grind all over to a depth of about 1 mm (1/32 in).

The result is a floor with a texture like sandpaper – a bit rougher than a burnished finish, but smoother than a float or broom finish. It is likely to have less wear resistance than a burnished floor, because its surface has not been densified by trowelling. Early-age grinding does a good job preparing a floor for coatings and coverings. The slightly rough texture makes for good adhesion, and the possible reduction in wear resistance hardly matters. There is debate, however, over the suitability of this finishing method for bare floors in warehouses and factories.

The normal tool for early-age grinding is a double-disc grinder with three abrasive blocks on each disc. The blocks should be made of 1.70-mm (US no. 12 sieve) grit. Though grinding takes place without added water, airborne dust is generally not a problem because the concrete is still moist.

The main benefit to early-age grinding – and the reason it was introduced – is the elimination of overtime work in cold weather. Instead of staying up all night to float and trowel a slab cast at low temperature, floorlayers go home after a normal workday and come back a day or two later to finish the job.

Never wildly popular, early-age grinding has fallen into almost total disuse. As far as I can tell, no one in the UK or America is actively promoting this method. Still, it remains an option worth considering where labour is dear or skill at trowelling scarce.

Dry shakes

Also called a sprinkle or dust finish, a dry shake consists of dry aggregate, coated with or mixed with cement, that is cast on the fresh concrete and worked in by floating. Depending on what aggregates they are made of, dry shakes can:

- increase a floor's resistance to wear;
- make a floor more slip-resistant;
- colour a floor;
- make a floor more electrically conductive.

Dry shakes are widely used, especially in America. In my view (with which many disagree), they are widely overused. Most dry shakes are proprietary products and heavily promoted. Without doubt some of them have value, but many are sold in situations where they do little good.

That is particularly true of wear-resistant dry shakes, which come in two basic types: non-metallic and metallic. The non-metallics consist of brittle mineral aggregate – that is, rock. The metallics consists of iron filings. Sadegzadeh and Kettle (1988), along with Scripture, Benedict and Bryant (1953), present persuasive evidence that non-metallic shakes have little or no effect on wear resistance. My own testing confirms that. The metallics, in contrast, clearly produce a marked improvement. But even metallics are often specified in locations where plain concrete, with a good finish, would serve as well.

Dry shakes have drawbacks that often go unrecognized. They often fail by delamination, with the shake peeling off in thin sheets. Good workmanship reduces the risk of delamination, but even the most careful contractors sometimes experience it. Another drawback is reduced flatness. The straightedge work that makes a floor very flat is not compatible with the application of a dry shake. Phelan (1989) suggests it is hard to achieve flatness better than $F_F 45$ on a dry-shake floor. However, the use of riding trowel machines with big float pans may be reducing the connection between dry shakes and poor flatness.

Designers should approach dry shakes as they should approach anything that raises the costs of a floor and introduces some risk: with caution and a "show me" attitude. If a dry shake meets a real need and does more effectively or cheaply than an alternative such as a topping or coating, use it. If it does not meet those conditions, spend the money elsewhere.

To apply a dry shake with success, requires both care and skill. Here are the typical steps:

1. Strike off normally and wait for bleed water to disappear.
2. Apply two thirds of the specified amount of dry shake material. Mechanical spreaders are widely used, but some floorlayers still throw it out with shovels.
3. Work the dry shake into the concrete with hand or power floats, or both.
4. Apply the remaining one third of the dry shake.
5. Again, work the dry shake into the concrete with floats.
6. Trowel as needed to give the desired finish.

Never use entrained air in concrete that will get a dry shake, because air greatly increases the risk of delamination.

Since the dry shake soaks up water from the concrete beneath it, that concrete must not be too dry. A dry-shake floor is not the place for stiff, low-slump concrete or low water–cement ratios. Slumps near 125 mm (5 in) are typical.

Vacuum dewatering

This is a technique for sucking water from fresh concrete. After the floor has been struck off, airtight mats are laid over it and vacuum is applied under the mats. Deacon (1986) suggests the vacuum should stay on for 3–5 minutes for each 25 mm (1 in) of slab thickness.

Vacuum dewatering eliminates bleed water and greatly reduces finishing time, especially in cool, damp conditions. You can usually start floating the floor right after the vacuum mats come off. As with early-age grinding, the chief benefit of vacuum dewatering is the elimination of late-night overtime in cold weather.

The process is said to improve concrete strength and wear resistance. Because the mats are about 5 m (16 ft) square, vacuum dewatering is best suited to narrow-strip pours. The technique is hard to apply to wide-strip and large-bay pours, though a few builders have done so.

Shot blasting

This abrades the floor surface with tiny steel balls. Its usual purpose is to prepare a floor for a bonded topping or coating (Chapters 18 and 22). Occasionally shot-blasting is used on a direct-finished concrete floor to improve its slip resistance. This usually comes about when a smooth trowel finish proves too slippery.

How to specify finishes

Many specifiers do a poor job on floor finishes. They rely on method specifications that try to tell the floorlayers what to do, step by step. The following language, which has appeared in hundreds of American specifications, is typical:

> *First, provide a floated finish. The finish shall next be power trowelled three times, and finally hand trowelled. The first trowelling after floating shall produce a smooth surface which is relatively free of defects but which may still show some trowel marks. Additional trowelling shall be done by hand after the surface has hardened sufficiently. The final trowelling shall be done when a ringing sound is produced as the trowel is moved over the surface.*

Reading that, I can't help imagining some earnest inspector cupping his ears, listening for that ringing sound. Whatever its merits as a general guide, a specification like that has at least two serious shortcomings when made part of the contract documents.

First, it clouds the issue of responsibility for poor results. If a builder follows every instruction to the letter, who is to blame if the finished floor has defects?

Second, it is counterproductive because if may force floorlayers to adopt unfamiliar methods. (Either that, or they ignore the instructions, creating a different sort of problem.) Good floorlayers develop their own method of working – a method suited to their own abilities and local condition. Two crews can achieve almost identical results using quite different methods.

Instead of telling floorlayers how to do their job, designers should describe the required finish and leave it to the floorlayers to sort out the methods. In other words, designers should write performance specifications rater than method specifications. For most uses, designers can control the floor finish by including these three items in their specifications:

- a requirement for surface regularity (see Chapter 19);
- a requirement for wear resistance (see Chapter 20);
- a description of the desired surface texture.

Floor surface regularity and wear resistance can be specified in terms of standard tests.

Surface texture is trickier because no standard test exists for it. All we can do is describe it, but in practice that works reasonably well. Most floorlayers know what is meant by "burnished finish", "float finish" and "broom finish".

18

Concrete Toppings

A concrete topping is a layer of concrete placed on top of a structural slab. Toppings are laid over both ground-supported and suspended floors. They fall into three main types: monolithic, bonded and unbonded (Figure 18.1).

Not every floor needs a topping. In the past, it was widely accepted that high-quality floors needed toppings, but few people argue that today. Well-made, direct-finished concrete slabs are as good as toppings for most purposes, and almost always cost less. The modern fashion is to use direct-finished concrete, with no topping, wherever possible.

Nevertheless, toppings have their place (Table 18.1). They are still used to:

- provide high resistance to wear;
- reduce material costs (but not labour) by putting strong concrete at the floor surface and weaker concrete in the base slab;
- provide a base for other construction work, with the topping laid after the other work has ended;
- to repair defects.

Monolithic toppings

Also called integral toppings, these are laid while the underlying concrete is still plastic. They are typically about 15 mm (5/8 in) thick, though there seems to be nothing special about that dimension. The coarse aggregate is usually no bigger than 10 mm (3/8 in).

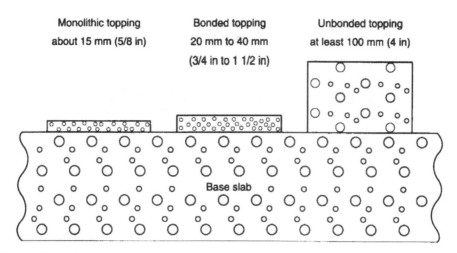

Figure 18.1 Topping thickness varies with type

Unless the base slab hardens or crusts over before the topping is laid, monolithic toppings usually live up to their name, with a strong bond between topping and base. For that reason, most designers count the topping as part of the overall floor thickness. If design calculations support a slab thickness of, say 150 mm (6 in), it would be normal to make the base slab only 135 mm (5-3/8 in) thick, making up the final 15 mm (5/8 in) in topping material.

The usual reason to choose a monolithic topping is to make the floor surface of stronger concrete than the mass of the slab. The topping concrete is normally a high-strength mix, sometimes containing special aggregates such as traprock. The base slab can then be made of ordinary, weaker concrete.

Another possible use of a monolithic topping is to create a coloured floor, without the cost of using pigment throughout a structural slab.

The drawback to a monolithic topping is higher labour cost, compared to a direct-finished slab. There is also a small risk of delamination if the builder lets the base slab sit for too long before laying the topping.

Table 18.1 Recommended thicknesses for concrete toppings

Topping Type	Thickness	
	Minimum	Maximum
Monolithic	10 mm (3/8 in)	20 mm (3/4 in)
Bonded	20 mm (3/4 in)	40 mm (1 1/2 in)
Unbonded	100 mm (4 in)	no limit

Bonded toppings

These are laid over a structural slab that has already hardened. Too thin to have much strength on its own, a bonded topping depends on its bond to the base slab. The coarse aggregate is usually 10 mm (3/8 in) – same as in monolithic construction.

BS 8204:Part 2 recommends that bonded toppings be 20–40 mm (3/4–1^1/$_2$ in) thick. It is risky to try a bonded topping much over 40 mm (1^1/$_2$ in) thick, because curling stresses are too likely to pull the topping off its base. If a topping must be over 40 mm (1^1/$_2$ in) thick, unbonded construction is the safer choice.

Because bonded toppings do not always stay bonded, structural engineers generally do not count them as part of the floor's structural thickness. There are exception, however. Some precast units for suspended floors rely on bonded toppings for structural strength.

Importance of bond

A bonded topping is nor better than its bond to the base slab. If bond fails, the topping is likely to break up under traffic.

You can test bond strength by isolating a small section of topping with a saw or core drill, gluing a ring to its surface and measuring the force needed to break it free. According to one expert in this field (Dennis, 1989), readings above 0.7 MPa (100 psi) denote satisfactory bond, while readings below 0.5 MPa (70 psi) reveal a problem. In contrast, concrete itself typically has tensile strength of 2–3 MPa (300–450 psi).

We could write performance specifications based on pull-off tests, but nobody I know does that. The normal approach is to specify the base-slab preparation and topping techniques known to work.

Preparing the base slab

Chapter 22 covers this subject, but a few points are worth noting here.

Though it's tempting to give the base slab a rough finish, you cannot count on that alone to provide a good bond. On the contrary, a rough finish may make matters worse because it traps dirt and is hard to clean. The safer course is to give the base a smoothish finish (not necessarily burnished), and count on roughening it by other means just before the topping goes on.

A recent and unusually thorough study (Well, Stark and Polyzois, 1999) suggests that among preparation methods shot-blasting provides the most reliable bond – so good that it eliminates any dependence on bonding agents. That same study shows that polymer-modified bonding agents worked no better than plain old cement–sand paste.

People may argue about preparation methods and bonding agents, but no one disputes the need for cleanliness. Even the thinnest layer of dust will inhibit bond. If you would not gladly eat a sandwich after dropping it on the base slab, that slab is not clean enough for a bonded topping.

Joints in bonded toppings

All movement joints in the base slab should extend upward through the topping. If they do not, the topping is almost sure to crack over every joint.

For our purposes here, movement joints include all isolation, expansion, contraction and warping joints. A construction joint may or may not be a movement joint, depending on the amount of reinforcement passing through it. Construction joints in suspended floors are usually not movement joints. If in doubt, assume the joint will move.

A topping can have more joints than its base slab. Extra joints are sometimes needed for construction convenience, or to divide the floor into narrow strips for better control over flatness.

Unbonded toppings

Also called overslabs, these are, in effect, separate slabs laid on top of the base slab. There is no attempt to bond the topping to the base. On the contrary, the topping performs more predictably if completely unbonded. Some designers cover the base slab with polyethylene sheets or building paper to prevent random bonding.

An unbonded topping should be at least 100 mm (4 in) thick. There is no maximum thickness, as long as the base slab can support the weight. Unlike the thinner monolithic and bonded toppings, it can be made with normal-size aggregates. Because it is unbonded, it can never be counted in the structural thickness of the slab.

For crack control, unbonded toppings are treated as ground-supported slabs (even if the base slab is suspended). Any of the crack-control methods discussed in Chapter 16 can be used. Most unbonded toppings contain light reinforcement and sawcut joints, but heavy reinforcement, steel fibres and post-tensioning are all possible.

One big advantage of unbonded construction is that the joints in the topping need not follow the joints in the base slab. There is no need to extend base-slab joints up into the topping, unless you expect very large movement.

Another advantage is that unbonded toppings can be laid over old surfaces that might not be suitable for bonding. An oil-contaminated slab makes a bonded topping almost impossible, but poses no problem in unbonded construction.

The chief drawback is that unbonded toppings raise the finished floor elevation by at least 100 mm (4 in). In new construction, you can anticipate this by depressing the base slab. It creates problems in remedial work, however. Ramps usually have to be built, and doorways may lack needed overhead clearance.

Forbidden thicknesses

Bonded toppings should not exceed 40 mm ($1\frac{1}{2}$ in) in thickness, lest curling stresses cause them to debond. Unbonded toppings should be at least 100 mm (4 in), lest they lack the structural strength to support traffic. That leaves us with forbidden range between 40 and 100 mm ($1\frac{1}{2}$ and 4 in).

What if the topping thickness needs to fall in the forbidden range? It happens more than you might suppose, thanks to the widespread use of wood-block flooring 60 mm ($2\frac{1}{2}$ in) thick. Many industrial buildings contain such flooring, and with changing uses owners often want to replace some of the blocks with concrete. The only topping that will not change the finish floor elevation is one that matches the thickness of the blocks.

Some designers have successfully bent the rules by adding steel fibres at about 40 kg/m^3 (70 lb/yd^3). That has not been done enough that we can consider it tried and true, but it's worth considering if a topping simply must fall in the forbidden thickness range.

Cement–sand screeds

The cement–sand screed is a special kind of topping made of cement, sand and just enough water to make the mix workable. There is no coarse aggregate.

Because screeds have poor resistance to wear, they are never used as a wearing surface. They are always covered with carpet, tile, vinyl sheets or similar materials. Screeds are often used in houses, offices, shops and institutions, but almost never in warehouses and factories.

Like regular concrete toppings, screeds can be either bonded or unbonded (Table 18.2). (There are no monolithic screeds, however.) The rules for bonding screeds are the same as those for bonding toppings. The base slab must be carefully prepared, and the screed should be 20–40 mm (3/4–1^1/$_2$ in) thick.

Unbonded screeds are similar to other unbonded toppings, but are sometimes thinner. While unbonded toppings made of ordinary concrete should be at least 100 mm (4 in) thick, unbonded screeds are sometimes only half as thick.

The floating screed is a special type of unbonded screed separated from the base slab by a layer of insulation (thermal or acoustic). Floating screeds should be at least 65 mm (2^1/$_2$ in) thick.

Screeds versus direct-finished slabs

Though useful, screeds are rarely essential. In almost every case where a screed could be considered, the alternative of a direct-finished slab also exists. The choice between a screed and a direct finish can be hard.

Designers choose screeds for these reasons:

- to improve surface regularity (flatness and levelness);
- to provide slopes for drainage;
- to hide a defective floor;
- to incorporate services such as wiring.

All those aims can also be met with direct-finished concrete. With care, a slab can be laid as flat and level as a screed, and can be sloped

Table 18.2 Recommended thicknesses for sand–cement screeds

Screed Type	Thickness	
	Minimum	Maximum
Bonded	20 mm (3/4 in)	40 mm (1^1/$_2$ in)
Unbonded	50 mm (2 in)	no limit
Floating	65 mm (2^1/$_2$ in)	no limit

for drainage. Defective concrete surfaces can be repaired by grinding and patching, instead of screeding. In-floor services provide perhaps the strongest argument for screeds, but even here there are alternatives.

Local fashion often determines the choice between screed and direct finish. If you build in a part of the world where screeds are common (much of Europe, for example), you can probably count on finding skilled screedlayers, but the local workers may lack the ability to put a good finish on a concrete slab. If you go to a region where screeds are rare (America, for example), you will have no trouble finding skilled concrete finishers, but screedlayers will be thin on the ground.

What about America?

Screeds are not much used in America, where the preference for direct-finished concrete slabs is all-powerful, even in residential and office work. But Americans might well rethink that preference.

The increasing focus on floor flatness has revealed that direct-finished suspended slabs are seldom very flat. They often lack the flatness needed to make a floor look good when covered with thin, shiny material such as sheet vinyl or vinyl composition tile (VCT). The result has been several serious disputes and many dissatisfied customers, most of them in the medical and educational fields.

If an American floor in a hospital or school is found not to be flat enough, the usual remedy is to cover all or part of it with a cementitious levelling compound that looks a lot like a sand–cement screed. That means many American floors are in fact getting screeds. They just were not designed that way.

Screed mixes

These are usually identified by ratio of cement to sand, by mass. Table 18.3 shows the two mix designs most common in the UK. The table shows US equivalents, but screeds are so rare in America that no standards for them exist. The 1:4 mix is for general use, and the 1:3 mix goes in high-traffic areas. Remember, though, that even a 1:3 screed needs a floorcovering. Murdock, Brook and Dewar (1991, pa. 405) recommend the stronger 1:3 mix for any screed that supports a thin covering such as carpet, sheet vinyl or vinyl tile, reserving the 1:4 mix for thick coverings such as ceramic tile.

Table 18.3 Screed mixes

Use	Mix by Dry Mass	Mix by Volume					
		UK			USA		
		Cement	Sand		Cement	Sand	
			Dry	Wet		Dry	Wet
Heavy traffic	1:3	1 bag	0.10m³	0.12m³	1 bag	3.0ft³	3.6ft³
General	1:4	1 bag	0.13m³	0.16m³	1 bag	3.9ft³	4.8ft³

The water content is typically 7–12% of the total mix, by mass. It is usually judged by eye. If you squeeze a handful of screed mix, it should clump together without any appreciable water being squeezed out.

Components can be batched by mass or volume. Batching by mass may be technically superior, but in practice most screedlayers batch by volume and that is normally good enough. When poor batching occurs – as it does all too often – the cause is more likely to be carelessness than any inaccuracy inherent in volume batching.

Because it is stiffer than the concrete used in floor slabs, screed material should be mixed in a forced-action mixer. Pan mixers and trough-type mortar mixers are suitable. Free-fall mixers are not. Ready-mix screed material, delivered by truck, is a dependable alternative to mixing on site but it is not available everywhere.

Screedlaying techniques

Over the years, screedlayers have developed techniques quite different to those used on concrete slabs and toppings.

One method is to lay the screed in strips about 3m (10ft) wide. Temporary forms – typically wood battens or steel angles – are fixed to the base slab. This method offers good control over flatness and levelness, but the joints sometimes show up as obvious bumps.

The other method is to lay the screed continuously, with no side forms. This is called wet screeding. It offers little control over levelness, but eliminate the construction joints.

The screed is compacted either with a roller or by tamping with a flat plate. The vibrators used to compact ordinary concrete do not work well on screeds. This step is important; many screed failures have been traced to poor compaction.

Finishing is a matter of smoothing the surface with straightedges, floats and trowels. Power tools are rarely used. There is no attempt to densify the surface with repeated trowellings, as is done with concrete.

Screeds should be cured, and the methods are the same as for concrete slabs.

Screed testing

Screed failures are both common and costly. They are common because screeds are often mixed on site and laid without much supervision or inspection. They are costly because they often go undetected till the floor is in use. Then the repair must deal not only with the screed, but also with its covering, which may cost more than screed and base slab combined. Timely tests can prevent most screed failures.

Effective testing must focus on the properties that matter. Some concrete tests have little meaning when applied to screeds. The slump test is useless on screed mixes. Strength testing is possible, but not very relevant because a screed mix is not a structural material.

Most screed failures involve debonding, poor surface regularity or local crushing. Turning that around, we can say the important properties are bond strength (only for bonded screeds), good surface regularity and resistance to local crushing. We can test screeds for all three properties.

Bond can be checked by tapping with a hammer or steel rod. A hollow sound denotes debonding. Bond strength can be quantified by pull-off tests.

Screeds can be tested for surface regularity using any of the methods described in Chapter 19. The straightedge test is the most common, but screeds can also be specified by F-number or by TR 34 category.

Resistance to local crushing is of special interest in screed construction. Regular concrete floors also need this property, but hardly ever lack it. In contrast, local crushing is a very common cause of failures in screeds. The problem shows up as a crumbling of the screed under a concentrated load such as a furniture leg or high-heeled shoe.

The UK's Building Research Establishment has developed a crushing test for screeds (Roberts, 1986). The BRE screed tester (Figure 18.2) consists of a 4-kg (9-lb) drop hammer that falls 1 m (39 in) on to a circular foot covering an area of 500 mm^2 (0.775 in^2). After four blows the depth of indentation is measured with a special gauge. Most screeds show test results between 0 and 5 mm (0 and 3/16 in). Table 18.4 shows

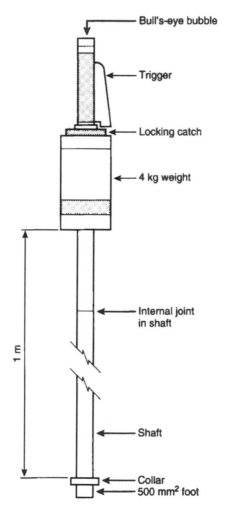

Figure 18.2 BRE screed tester

recommended values. The test is suitable for most screeds, but not for floating screeds or those made with lightweight aggregate.

An older method, the Stanger iron-nail test – is also used to check screeds for resistance to crushing. The inspector tries to hammer a $2^{1}/_{2}$-in, 10-gauge wire nail into the screed. If the nail goes in without bending (a steady hand helps), the screed is unacceptable. It the nail bends, a second nail is applied at the same spot. If it too bends, the screed is good. If the second nail penetrates, the screed is questionable. One good feature of the Stanger test is that, unlike the BRE screed tester, it does not require a bare screed. You can drive the nail right through carpet and some other floorcoverings.

Table 18.4 Classification of cement–sand screeds by results of BRE screed test

		Maximum Allowable Indentation From 4 Blows	
Category	Use	Random Tests[1]	Grid Tests[2]
A	Areas of very heavy traffic, areas where later disruption would be unacceptable, dust-free rooms	2 mm	3 mm
B	Areas of heavy traffic such as corridors, lobbies, shops, and restaurants	3 mm	4 mm
C	Areas of light traffic such as offices and houses	4 mm	5 mm

[1] In random testing, there is about one test for every 10 m² (100 ft²), with special attention to vulnerable areas such as doorways and joints.
[2] In grid testing, tests are made at points 2 m (6 ft) apart in both directions.
In either case, 10% of the readings are allowed to exceed the specified maximum by up to 1 mm.

Terrazzo

This is a special type of bonded concrete topping used for its good looks. Terrazzo toppings are traditionally laid in small panels about 1 m (3 ft) square, separated by thin strips of metal or hard plastic. After the concrete has hardened, it is ground smooth and polished with high-speed machines.

Terrazzo is concrete, but its ingredients are chosen with appearance in mind. The aggregate consists of marble chips, ranging in size from 2 to 25 mm (1/16 to 1 in). The cement is usually white Portland, but ordinary grey cement is also used. Pigments are sometimes added. The mixes are always cement-rich, with a 1:2 or 1:2.5 ratio of cement to marble aggregate. Most terrazzo contractors mix their material on site.

While Portland cement (white or otherwise) is the traditional binder, nowadays much terrazzo is made with epoxy or other polymers.

Terrazzo is struck off, floated and trowelled in much the same way as other concrete toppings. It does not call for great care in trowelling, however, because the ultimate finish comes not from the trowel but from the grinders.

Grinding typically takes place about two days later. The first grinding pass uses coarse abrasive blocks of No. 20 grit. Later passes take progressively finer blocks. The final polish comes from No. 60 or 80 grit abrasive.

Terrazzo can be laid directly over the concrete slab, but some specialists prefer to lay it over a sand–cement screed. This creates a three-course floor, complicating the job but saving money on costly terrazzo material.

Terrazzo is a good choice for public areas where good appearance must be coupled with resistance to heavy foot traffic. It lasts a long time and is easy to clean. It is often seen in building entrances, airports and railway stations. Though not suitable for industrial traffic, it holds up well under light vehicles such as luggage trolleys.

Summary – choosing a topping

No one type of topping is best for every job.

Monolithic toppings cost less than other types and add to the slab's structural thickness. But they must be laid at the same time as the base slab.

Bonded toppings can be installed at any time, but require great care to prevent debonding. They often cost more than rival types.

Unbonded toppings can also be installed at any time. They often cost less than bonded toppings, though more than monolithic. They come into their own over old floors because they can be laid without regard for joints and cracks in the base slab. Their main drawback is the effect on finished floor elevation, which must rise by at least 100 mm (4 in).

Cement–sand screeds are useable only where the floor will be concealed by carpet, tile or another floorcovering.

Terrazzo toppings are used mainly for visual effect. They provide a good-looking floor surface that resists heavy foot traffic.

Lastly, *direct-finished concrete slabs* are almost always an alternative to toppings. They usually cost less.

19

Surface Regularity

Surface regularity has been the focus of great interest in recent years. For most of the twentieth century, floor designers either ignored surface regularity or tried to control it with specifications based on a 3-m (10-ft) straightedge. Those straightedge tolerances were rarely enforced; indeed, effective enforcement was next to impossible because there was no standard test.

So the situation might have remained, except for a development in the materials-handling field. In the early 1970s, forklift makers introduced the very-narrow-aisle (VNA) turret truck. It differed from ordinary forklifts in having its fork mounted on a rotating turret. While ordinary forklifts needed aisles about 4 m (13 ft) wide, the new turret trucks could run in aisles only 2 m (6 ft) wide. Narrower aisles meant greater storage density – something many warehouse users sought.

The turret truck had one glaring drawback, however. It needed a flat and level floor. Warehouse users who tried to run turret trucks on ordinary 1970s floors soon learned that neither the 3-m (10-ft) straightedge nor the floorlaying methods then in use were good enough for the new vehicles.

To satisfy turret-truck users, floor designers and builders developed ways to build flatter and more level concrete slabs. Their efforts led to the superflat floor, discussed at the end of this chapter.

Perhaps even more importantly, the sharper focus on surface regularity led people to question the flatness and levelness of all floors, not just those that supported turret trucks and needed a superflat surface. The result was radical new ways to specify and test floor surface regularity: the F-number system in America and the TR 34 system in the UK.

But before we look at those systems, we need to consider two key distinctions: between flatness and levelness, and between random and defined traffic.

Flatness and levelness

These terms have overlapping definitions in everyday speech. But within the field of concrete floors, flatness and levelness mean different things.

Flatness is defined as planarity or lack of curvature. A flat floor is smooth and free of bumps or dips. An unflat floor is bumpy or wavy.

Levelness is defined as horizontality or lack of slope. A level floor is horizontal. An unlevel floor is sloped or tilted.

A few examples may help illustrate the distinction between flatness and levelness. In normal use, a billiards table is both flat and level. If we put bricks under one end, the table will still be flat but it will no longer be level. On a calm day, the surface of a pond is both flat and level. If the wind rises the surface will cease to be flat, but it remain level.

Some floors are designed to be unlevel. They are sloped for drainage or to minimize site grading. I have never hear of any floors designed to be unflat, though many turn out that way.

Flatness and levelness are both desirable properties (except on those floor with design slopes), but they have different implications for the floor user. Flatness is critical where the main issue is the behavior of wheeled vehicles. Flatness is also important where appearance matters – as in an institutional floor that will be covered with vinyl tile. In contrast, levelness is critical where the main issue is the installation of fixed equipment such as warehouse racks or machine tools.

While exceptions exist, most users find flatness matters more than levelness. If a floor is unlevel, fixed equipment can be shimmed or jacked up to accommodate it. But it is not so easy to adapt a wheeled vehicle to a surface that is not flat enough for it.

Flatness and levelness also have different implications for the floorlayer. Flatness comes from finishing; it is determined mainly by the finishing methods, and to a lesser extent by the finishability of the concrete mix. Levelness comes from forms and strike-off; it is determined mainly by the accuracy of the side forms and the method used to strike of the concrete.

Defined versus random traffic

Floors get two kinds of traffic: defined and random. On a defined-traffic floor, vehicles are confined to fixed paths. On a random-traffic floor, vehicles (or walkers) are free to roam, though inevitably some traffic paths see more use than others.

The distinction matters because the two kinds of traffic call for different ways of measuring surface regularity. On a defined-traffic floor, you can measure a continuous, or nearly continuous, profile down every travel path. You can check effectively all of the floor profile that matters to the vehicle user. In contrast, a random-traffic floor has an infinite number of possible travel paths. You could never measure them all, so you have to fall back on statistical sampling. You check selected points or lines and hope they represent the whole surface.

Historically, the highest requirements for surface regularity have been found in the defined-traffic category. The reason is that turret trucks – the vehicles that demand the flattest and most level floors – run in defined paths. It would be wrong, however, to assume that defined-traffic floors are always superior in surface regularity.

Though defined-traffic floors are important, their random-traffic cousins are far more numerous. Defined-traffic floors are found in very-narrow-aisle warehouses and in buildings with automatically-guided vehicles – and almost nowhere else. The great majority of industrial floors, as well as all non-industrial floors, fall in the random-traffic category.

F-numbers

Since its adoption into ACI and ASTM standards in the late 1980s, the F-number system has become the most common way to specify and measure floor surface regularity in the United States and Canada. It is also widely used in Latin America and Australia. Europeans, in contrast, are more likely to use the TR 34 system or the German standards DIN 18 202 and DIN 15 185.

The F-number system includes three separate numbers:

- F_{min}, which encompasses both flatness and levelness and is used only on defined-traffic floors;
- F_F, which controls flatness and is used only on random-traffic floors;
- F_L, which controls levelness and is used only on random-traffic floors.

Defined-traffic floors are specified by the single number F_{min}. Random-traffic floors are normally specified by a pair of numbers, written as F_F/F_L. However, some suspended random-traffic floors are specified by F_F alone.

The three numbers have this much in common:

- A higher F-number always denotes a better floor.
- The normal scale runs from 10 to 100. Only the worst floors exhibit F-numbers below 10, and only a few of the best measure above 100.
- The scales are linear. If you double the F-number, you halve the magnitude of the allowed deviations.

F_{min} for defined traffic floors

The F-number system controls defined-traffic floors by means of a single number called F_{min}. F_{min} defines both flatness and levelness, and is measured over the wheel spacings of the vehicle that will travel on the floor. Testing is confined to the travel paths.

When you say a floor measures, say, $F_{min}50$, you are not making a general statement about the floor's flatness and levelness. You are saying that when a particular vehicle travels in a particular path, it will not encounter any slope or change in slope greater than that allowed by the F_{min} equations, shown below:

In SI units:

$$d_{max} = \frac{[6.55(L+69)^{0.5} - 48]}{F_{min}}$$

$$e_{max} = \frac{33.3}{F_{min}}$$

where:

d_{max} = maximum elevation difference between points separated by L, in mm;

e_{max} = maximum rate at which elevation difference between points separated by L can change with travel across the floor, in mm/100mm;

L = distance between wheels (transverse) or axles (longitudinal), in mm;

F_{min} = the defined-traffic F-number (dimensionless).

In fps units:

$$d_{max} = \frac{[1.3(L+2.7)^{0.5} - 1.9]}{F_{min}}$$

$$e_{max} = \frac{4.00}{F_{min}}$$

where:

d_{max} = maximum elevation difference between points separated by L, in inches;

e_{max} = maximum rate at which elevation difference between points separated by L can change with travel across the floor, in inches/foot;

L = distance between wheels (transverse) or axles (longitudinal), in inches;

F_{min} = the defined-traffic F-number (dimensionless).

The tolerances d_{max} and e_{max} are usually expressed to 0.1 mm or 0.001 in. The equations for d_{max} (levelness) are good for wheel or axle spacings from 300 mm to 3 m (1 to 10 ft). The equations for e_{max} (flatness) are good for wheel or axle spacings from 1 to 3 m (40 in to 10 ft). Almost all vehicles used on defined-traffic floors have dimensions within those ranges.

F_{min} is normally measured with a differential profileograph set up to match the wheel and axle spacings of the defined-traffic vehicle. The profileograph measures the elevation difference between left and right wheels (the transverse-differential profile) and between front and rear axles (the longitudinal-differential profile). Both profiles are checked for compliance with the calculated limits for d_{max} and e_{max}.

Specified F_{min} values range from 40 (rarely lower) to 100 (never higher). F_{min}100 is the traditional superflat specification, and is reserved for VNA warehouses with very tall racks or very high throughput, or both. In recent years, F_{min}60 has become popular for less-demanding VNA applications. With laser screeding, contractors can lay F_{min}60 floors in large bays, eliminating the costly narrow-strip layout needed for F_{min}100. F_{min} 40 is sometimes specified for the least demanding VNA applications, and also for low-rise automatically-guided vehicles.

F_F/F_L for random-traffic floors

The F-number system controls random-traffic floors with a pair of numbers, F_F and F_L. The flatness number, F_F, is based on the floor curvature over 600 mm (24 in) (see Figure 19.1). The levelness number, F_L, is based on the floor slope over 3 m (10 ft) (see Figure 19.2).

ASTM E 1155 (fps units) and ASTM E 1155M (SI units) describe the standard test for F_F and F_L. A typical survey involves these steps:

1. Determine the minimum number of 3-m (10-ft) slope readings. For test sections of 150 m² or more, the minimum is the area divided by 3. (In fps units, it's the area in ft² divided by 30, for sections of 1600 ft² or more.)
2. Lay out enough straight lines on the floor to provide at least the minimum number of 3-m (10-ft) slope readings. The number of readings from each line is equal to the length in m, minus 2.7 m, divided

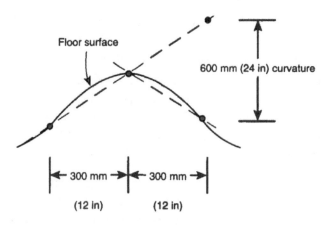

Figure 19.1 600-mm (24-in) curvature determines the flatness number, F_F. TR 34 calls this Property II

by 0.3 m. (In fps units, the number of readings from each line is simply the length in ft, minus 9 ft.) Figure 19.3 shows one possible layout, with all lines oriented 45° to the test section's longest side. It is also permissible, where the test section is at least 7.5 m (25 ft) wide, to lay out runs parallel and perpendicular to the longest side. Under most circumstances, survey lines must avoid slab edges, construction joints and penetrations by at least 600 mm (2 ft).
3. Measure point elevations on 300-mm (12-in) centres down each survey line.
4. Calculate the elevation difference between each pair of adjacent points. This value is the 300-mm (12-in) slope.
5. Calculate the difference between each pair of consecutive 300-mm (12-in) slope readings. *This is the 600-mm (24-in) curvature.*

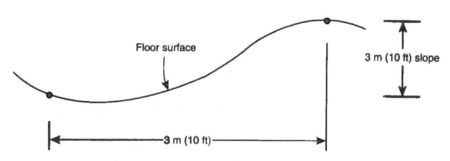

Figure 19.2 3-m (10-ft) slope determines the levelness number, F_L. TR 34 calls this Property IV

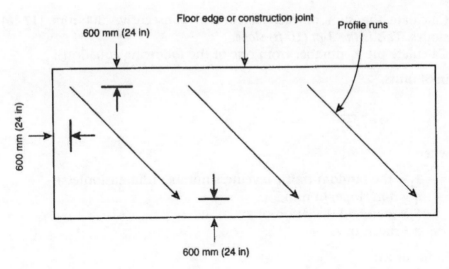

Figure 19.3 One way to lay out an F-number survey is with all lines at 45° to the slab's longest side

6. Calculate the F_F number from one of the following equations:

   In SI units:

   $$F_F = \frac{114}{(3S_q + q)}$$

   where:

   F_F = the random-traffic flatness number (dimensionless);
   q = 600-mm curvature, in mm;
   S_q = standard deviation of q;
   q = mean of q.

   In fps units:

   $$F_F = \frac{4.57}{(3S_q + q)}$$

   where:

   F_F = the random-traffic flatness number (dimensionless);
   q = 24-in curvature, in inches;
   S_q = standard deviation of q;
   q = mean of q.

7. Calculate the sum of each set of ten consecutive 300-mm (12-in) slopes. *This is the 3-m (10-ft) slope.*
8. Calculate the F_L number from one of the following equations:

In SI units:

$$F_L = \frac{315}{(3S_z + z)}$$

where:

F_L = the random-traffic levelness number (dimensionless);
q = 3-m slope, in mm;
S_z = standard deviation of z;
z = mean of z.

In fps units:

$$F_L = \frac{12.5}{(3S_z + z)}$$

where:

F_L = the random-traffic levelness number (dimensionless);
z = 10-ft slope;
S_z = standard deviation of z;
z = mean of z.

ASTM standards allow a variety of tools for measuring F_F and F_L, including optical levels, lasers and levelled straightedges with depth gauges. However, almost everybody uses an electronic inclinometer – either the walking type with feet 300 mm (12 in) apart (the Dipstick is the best-known brand name), or the rolling type with wheels separated by the same distance. Figure 11.11 in Chapter 11 shows an electronic inclinometer.

The resulting F-numbers will be practically the same whether surveyed in SI or fps units.

While no fixed relationship exists between F_F and F_L, the former averages about 40% higher than the latter on ground-supported slabs. On suspended floors the ratio of F_F to F_L tends to run higher.

ACI 117 recommends that a two-tiered format for random-traffic F-number specifications. The *overall* F-numbers apply to the floor taken as a whole. The *minimum local* F-numbers apply to smaller sections of the floor. Minimum local number are always lower than the overall numbers – typically one third to one half lower.

The two-tiered format, though sometimes criticized as needlessly complex, has proven an effective tool for improving floor flatness and levelness.

It pushes builders toward higher numbers, without punishing them brutally for the occasional bad day. The goal is to lay the whole floor to the overall F-numbers. But if the floorlayers do worse than that on an individual slab, or part of a slab, it can still be accepted provided it meets the minimum local F-numbers. Such an occurrence serves as a warning that the crew must do better on later slabs, to bring the overall numbers up to the specified values. If a test section fails to meet even the minimum local values, it must be repaired or replaced. But such failures are rare if the number are chosen with care and if all parties understand the rules from the start.

Table 19.1 shows overall and minimum local F-numbers for four classes of surface regularity. The classification comes from ACI 117, but the recommended applications are mine.

Construction joints

ASTM E 1155 and ASTM E 1155M forbid measuring F_F and F_L within 600 mm (24 in) of any slab edge or construction joint. (There is an exception for very small slabs.) Because the edges and construction joints are statistical anomalies, including them in the calculations would distort the results.

Table 19.1 Classification of floor flatness and levelness by F-numbers – for random-traffic floors (ACI 117)

		Minimum F-numbers required	
Floor class	Application	Overall	Local
Very flat	Warehouses with reach trucks over 8 m (26 ft) high, random-traffic areas in some VNA warehouses, air-bearing systems	F_F50/F_L30	F_F25/F_L15
Flat	Offices with moveable partitions, warehouses with mobile racks or reach trucks up to 8 m (26 ft high), busy general-purpose warehouses	F_F30/F_L20	F_F15/F_L10
Conventional Straightedged	Warehouses and factories where surface regularity is not critical	F_F20/F_L15	F_F15/F_L10
Bull-floated	Foot traffic, garages	F_F15/F_L13	F_F13/F_L10

The exclusion of construction joints from testing causes some problems, because the areas adjacent to construction joints are often – I would say usually – the least flat parts of a floor. It is hard to justify a system that excludes the spots where bad readings are most likely to occur.

The 1996 editions of ASTM E 1155 and ASTM E 1155M offer a solution. Designers can specify a limit for 600-mm (12-in) curvature at construction joints. The usual tolerance is 3.0 or 4.0 mm (0.125 or 0.150 in). The standard test calls for at least one test for each 3-m (10-ft) length of joint. Enforcement is normally limited to those construction joints that will be subject to traffic.

Suspended floors

Random-traffic F-numbers apply to suspended floors, but with some restrictions.

American standards let us enforce the flatness number, F_F, on any type of suspended floor. We need to recognize, however, that F_F values above 30 can be very hard to achieve on suspended slabs.

While making a suspended slab flat is hard, making it level is even harder. In view of that fact, ACI 117, the main American standard for concrete tolerances, prohibits the enforcement of F_L tolerances on suspended floors, with the exception of fully shored slabs that are tested while the shores are still in place. Unshored floors, including most slabs on metal deck, are excluded from levelness testing.

If a suspended floor needs to be very flat or very level – luckily most do not – the best choice is usually a topping or screed.

The TR 34 system

The Concrete Society's Technical Report 34, *Concrete Industrial Ground Floors*, commonly called just *TR 34*, includes a thorough, well-thought-out system for specifying floor flatness and levelness on industrial floors. The TR 34 system shares many features with F-numbers – a fact that stems from their common origin in research carried out by Allen Face in the early 1980s.

Like F-numbers, TR 34 controls both flatness and levelness. Again like F-numbers, TR 34 has different specifications for defined traffic and random traffic, which TR 34 calls "defined movement" and "free movement".

Perhaps the most significant difference between F-numbers and TR 34 lies in the way tolerances are presented. F-numbers are continuously variable.

The designer must choose a number and can, within practical limits, choose any number. While ACI 117 does classify floors (see Table 19.1), designers remain free to specify other values, and many do so. TR 34, in contrast, divides floors into a small number of standard categories, each having its own set of tolerances. The designer does not set a tolerance, but only picks a category. I confess a slight prejudice in favour of the F-number approach, while recognizing that standard categories have merit, too.

TR 34 properties

At the heart of the TR 34 system are four properties, identified by Roman numerals I–IV.

Property I (see Figure 19.4) is the elevation difference between two points separated by 300 mm (12 in). This is a levelness property. It applies only to defined-traffic floors, where it is measured parallel to the direction of travel.

Property II (see Figure 19.1) is the arithmetic difference between a pair of consecutive Property I readings. This is a flatness property. It applies both to defined-traffic and random-traffic floors. On defined-traffic floors it is measured parallel to the direction of travel. On random-traffic floors it is measured in two directions. Property II is the same as 600-mm (24-in) curvature in the F-number system.

Property III (see Figure 19.5) is the elevation difference between the left and right wheelpaths in which a vehicle runs. This is a levelness property. It applies only to defined-traffic floors, where it is measured in the direction perpendicular to travel, with readings taken on 300-mm (12-in) centres. It is the same as transverse levelness in the F-number system.

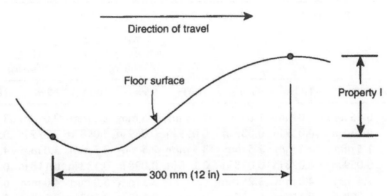

Figure 19.4 TR 34 Property I

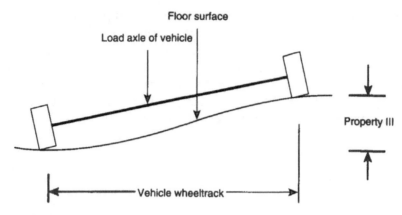

Figure 19.5 TR 34 Property III

Property IV (see Figure 19.2) is the elevation difference between two points separated by 3 m (10 ft). This is a levelness property. It applies only to random-traffic floors, where it is measured on a 3-m (10-ft) grid. It is the same as 3-m (10-ft) slope in the F-number system.

TR 34 for defined-traffic floors

TR 34 divides defined-traffic floors into three categories of surface regularity, called Superflat, Category 1 and Category 2. Each category is associated with limits on Properties I, II and III (see Table 19.2).

Table 19.2 TR 34 tolerances for defined-traffic floors

	Tolerance								
	Property I		Property II		Property III				
					Wheeltrack up to 1.5 m (60 in)		Up to 1.5 m (60 in)		
Category	95%	100%	95%	100%	95%	100%	95%	100%	
Superflat	0.75 mm	1.0 mm	1.0 mm	1.5 mm	1.5 mm	2.5 mm	2.0 mm	3.0 mm	
	0.030 in	0.039 in	0.039 in	0.059 in	0.059 in	0.098 in	0.079 in	0.118 in	
1	1.5 mm	2.5 mm	2.5 mm	3.5 mm	2.5 mm	3.5 mm	3.0 mm	4.5 mm	
	0.059 in	0.098 in	0.098 in	0.138 in	0.098 in	0.138 in	0.118 in	0.177 in	
2	2.5 mm	4.0 mm	3.25 mm	5.0 mm	3.5 mm	5.0 mm	4.0 mm	6.0 mm	
	0.098 in	0.157 in	0.126 in	0.197 in	0.138 in	0.197 in	0.157 in	0.236 in	

Note that TR 34 imposes two limits on each property. No readings are allowed to exceed the 100% limit, but up to 5% of the readings may exceed the 95% limit. The use of two limits represents an attempt to control not just the extreme values, but also the distribution of readings below those extremes.

TR 34 ties the recommended category to the lift height of the truck, as follows:

- Superflat for lift height over 13 m (43 ft);
- Category 1 for lift height of 8–13 m (26–43 ft);
- Category 2 for lift height under 8 m (26 ft).

That's useful, but it oversimplifies the situation. The need for surface regularity depends not only on lift height, but also on vehicle speed, vehicle type and the overall demands of the floor user.

Since the TR 34 properties are based on readings taken at 300-mm (12-in) intervals, floors specified in this way can be tested with same electronic inclinometers used to measure random-traffic floors in the F-number system. In practice, however, everybody uses differential profileographs like those used to measure F_{min}. But while an F_{min} profileograph is set up to match the wheel pattern of the vehicle, a TR 34 profileograph is set up to match just one axle of the vehicle (normally the wider of the two axles, if they differ). This takes care of Property III. Trailing wheels 300 mm (12 in) behind the axle allow measurement of Properties I and II.

All this may be changing, however. The latest version of TR 34 (Concrete Society, 2003) presents, in Appendix C, an alternative method for defined-traffic floors. The alternative method is much closer to the way defined-traffic floors are measured in the F-number system. It relies on full wheel-pattern testing, with tolerances adjusted for the distance between wheels. Floors are divided into three categories, called DM (for defined-movement) 1, 2 and 3. The classes correspond roughly to the current categories Superflat, 1 and 2.

TR 34 for random-traffic floors

TR 34 divides random-traffic floors into three categories of surface regularity, called FM (for free-movement) 1, 2 and 3. Each category is associated with limits on Properties II and IV. (see Table 19.3). As with the defined-traffic floors, each Property is subject to both a 95% and a 100% limit. For a floor to meet its specification, no more than 5% of the readings can exceed the 95% limits, and no readings can exceed the 100% limits.

Table 19.3 TR 34 tolerances for random-traffic floors

| | | Tolerance | | | |
| | | Property II | | Property IV | |
Category	Application	95%	100%	95%	100%
FM 1	Where very high standards of flatness and levelness are required	2.5 mm 0.098 in	4.0 mm 0.157 in	4.5 mm 0.177 in	7.0 mm 0.276 in
FM 2	Warehouses with stacking over 8 m (26 ft) high, transfer areas in VNA warehouses	3.5 mm 0.138 in	5.5 mm 0.217 in	8.0 mm 0.315 in	12.0 mm 0.472 in
FM 3	Warehouses with stacking up to 8 m (26 ft) high, factories, retail stores	5.0 mm 0.197 in	7.5 mm 0.295 in	10.0 mm 0.394 in	15.0 mm 0.591 in

Despite the fact that Properties II and IV are identical to the properties that determine F_F and F_L in the F-number system, they are not surveyed in exactly the same way. TR 34 calls for testing random-traffic floors on a 3-m grid. Property IV is measured between the grid points. Property II is measured along the some of the lines that connect the grid points, chosen so the total line length in metres equals or exceeds one tenth the floor's area in square metres. Unlike an F_F/F_L survey, a TR 34 random-traffic survey normally calls for two separate instruments: an optical level to check Property IV, and a rolling or walking gauge to check Property II.

Despite the differences in testing, we can still reasonably compare the FM categories to F-numbers (see Table 19.4). Interestingly, British and American specifications are fairly close on flatness but differ substantially on levelness. TR 34 imposes levelness requirements that American floorlayers would consider onerous. One reason for this difference may be that TR 34 explicitly addresses industrial floors, which are usually ground-supported and are often laid to high standards. American standards, in contrast, deal with a wider range of floors, including suspended slabs.

Table 19.4 Random-traffic specifications compared

TR 34 category	Equivalent F-numbers
FM 1	$F_F 30/F_L 46$
FM 2	$F_F 22/F_L 26$
FM 3	$F_F 16/F_L 21$

TR 34 on overall elevation

Properties I–IV do not directly control the floor's overall elevation. For most uses, overall elevation matters less than local flatness and levelness – but that is not to say it does not matter at all. TR 34 recommends an overall tolerance of ±15 mm (±5/8 in) from datum, for all floor classes.

Straightedge tolerances

Some specifiers continue to specify floor flatness by reference to a 3-m (10-ft) straightedge (see Figure 19.6). This method has two serious drawbacks, however.

The first is the so-called washboard loophole (see Figure 19.7). Checked with a straightedge both profiles in Figure 19.7 are equal. Yet for almost any use, the profile on top is clearly the better of the two.

The second drawback is the lack of a standard test. While ACI 117 and BS 8204:Part 2 both offer a bit of guidance in straightedge use, neither comes close to providing a complete test method like those available for F-numbers and TR 34 tolerances. No standard answers these questions:

- Where is the floor measured?
- How many tests need to be made?
- Are any failures allowed – and if so, how many?

Without answers to those questions, anyone trying to enforce a straightedge specification is likely to face a dispute.

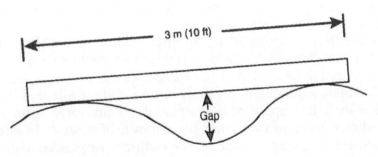

Figure 19.6 The straightedge test

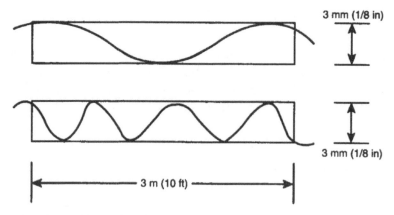

Figure 19.7 The washboard loophole

Factors that affect surface regularity

These are the most important factors:
- structural deflection;
- curling;
- width of pour;
- accuracy of side forms;
- striking off and finishing.

Structural deflection

Suspended floors all deflect, and more than a few deflect excessively. Deflection is hard to control and predict.

Deflection has a huge effect on levelness, and a smaller but still significant effect on flatness. Deflection is a serious problem for suspended floors that need to be level. ACI 117 recognizes this by simply eliminating the F_L levelness requirement on most suspended floors. It says "levelness tolerance shall not apply to slabs placed on unshored form surfaces and/or shored form surfaces after the removal of shores". In effect, the ACI standard is telling us not to use ordinary suspended slabs where levelness matters.

That is good advice, to the extent we can follow it. But some suspended floors do need to be level, no matter what ACI 117 tells us. What are our options?

One answer is to reduce the deflections. Chapter 6 contains some ideas on that. Another answer is to lay a thin topping over the suspended structural slab. But a topping won't stop the floor deflecting under loads applied later.

Deflection is seldom a problem on ground-supported floors.

Curling

This is a major cause of poor surface regularity on ground-supported floors, concrete toppings and slabs cast on metal deck. Testing flatness and levelness during construction usually fails to detect curling, because the floor curls afterwards.

Chapter 14 discusses curling and presents some ways to minimize its effect on floor flatness and levelness.

Width of pour

The narrower the slab, the easier is it to make the floor flat and level. The main reason for this is that a narrow pour allows a more accurate strike-off, at least with traditional (before the laser screed) methods.

Superflat floors are almost always laid in narrow strips no more than 6 m (20 ft wide).

In wide-strip construction, where the floor is struck off with a vibrating screed, floor levelness tends to deteriorate as the pours get wider. Some vibrating screeds can span 24 m (80 ft), but you cannot expect them to maintain a very level surface over that distance.

Laser screeding reduces the connection between slab width and surface regularity. With a laser screed, some floorlayers can achieve good flatness and levelness on pour as much as 30 m (100 ft) wide.

Accuracy of side forms

Side forms matter because they determine the elevation to which the floor is struck off. Forms have a big effect on levelness, but much less effect on flatness.

Laser screeding does not eliminate the importance of accurate side forms. Though the laser screed itself does not rely on the forms, its action stops about 0.5 m (2–3 ft) from the slab edges. The accuracy of the forms determines the surface regularity of that narrow strips between the slab edge and the laser-screeded surface.

With ordinary levels, it is hard to set forms closer than ±3 mm (±1/8 in). But that is good enough for most floors.

Superflat floors need better forms. Some floorlayers who specialize in superflat work use special forms with an adjustable top edge. Others use what look like ordinary steel or timber forms, but they take extra care in setting them.

It is not enough to set the forms level. They need to stay level till the concrete is finished. Forms need strong bracing to stand up to a heavy vibrating screed.

Striking off and finishing

The method of strike-off has a big effect on levelness, but only a smaller effect on flatness. Finishing methods work the other way around, heavily affecting flatness but doing little for levelness.

Here are the main tools used for striking off, listed in order of decreasing levelness achieved:

- hand straightedge riding on forms (narrow strips only);
- full-size laser screed;
- vibrating screed on narrow strips;
- roller screed;
- vibrating screed on wide strips;
- lightweight laser screed;
- hand straightedge on wet screeds.

When it comes to finishing, the main things to consider are the amount of straightedge work and the use of pans for floating. Working the floor with hand straightedges, including highway straightedges, dramatically improves flatness. Floorlayers aiming for superflat tolerances normally straightedge the concrete many times in two directions, scraping and filling as the concrete grows harder. The narrow-strip layout almost universal for superflat floors makes the straightedge work more convenient.

Straightedges are less effective on wide-strip and large-bay pours. There floorlayers aiming for tight tolerances often choose pans for floating.

Pans – especially the big ones used with riding trowel machines – flatten a floor better than float shoes, and far better than hand floating.

Dry shakes tend to make floors less flat (but not less level) because they interfere with the finishing steps that enhance flatness. Phelan (1989) suggests that we avoid dry shakes if the required degree of flatness is higher than $F_F 45$.

Superflat floors

There is no universally accepted definition, but in general a superflat floor is one near the upper limit of the achievable. In the TR 34 system, the term superflat comes with a specific set of tolerances. American usage tends to be looser, but by anyone's standards an $F_{min} 100$ floor would be considered superflat. In any case we are talking about a floor well over twice as flat and level as the typical concrete slab.

Figure 19.8 Superflat construction – filling a low spot that the highway straightedge has revealed

330 Design and Construction of Concrete Floors

The first recorded superflat floor – for Boots in Nottingham, England – was laid in very narrow strips (about 2 m or 6 ft wide), with screw-adjustable forms.

Modern practice is much closer to normal narrow-strip construction, but with some extra precision and straightedge work. These steps are typical:

1. Lay out the floor in strips no more than 6 m (20 ft) wide, with construction joints in non-critical areas.
2. Set forms as accurately as possible with an optical level.

Figure 19.9 Superflat construction – marking a new slab for grinding, after testing with a profileograph

3. Check forms with an electronic inclinometer (see Figure 11.11 in Chapter 11).
4. Re-check forms just before pouring concrete and adjust as needed.
5. Pour concrete at a slump of 100–125 mm (4–5 in) with minimal variation between batches.
6. Compact concrete with poker vibrators.
7. Strike off with a hand straightedge riding on the forms.
8. Strike off again with a hand straightedge riding on the forms.
9. Strike off with a 3-m (10-ft) highway straightedge at right angles to the forms. Lap each pass by 1.5 m (5 ft).
10. Strike off third time with a hand straightedge riding on the forms.
11. Strike off again with a 3-m (10-ft) highway straightedge.
12. Wait for bleed water (if any) to form and then disappear.
13. Power float, making passes across the slab width.
14. Right after floating, scrape with a 3-m (10-ft) highway straightedge. Fill low spots with material scraped from high spots (see Figure 19.8).
15. Wait as long as possible.
16. Power trowel.
17. If possible, scrape again with a 3-m (10-ft) highway straightedge. Filling low spots will probably not be possible at this stage, but scraping high spots may be.
18. Continue power trowelling to produce the desired finish (usually burnished).
19. Survey the floor and mark out-of-tolerance spots (see Figure 19.9).
20. Grind minor defects with a 250-mm (10-in) diamond disc.

As with any list of floorlaying steps, this one should be regarded as descriptive, not prescriptive. Some floorlayers get good results from somewhat different methods. On the other hand, following these steps to the letter will not guarantee a superflat floor.

20

Resistance to Wear

Wear is the erosion of a floor surface caused by solid objects rolling or sliding on it. Wear occurs every time a vehicle travels over a floor. Even foot traffic causes wear, as you can see on the steps of building buildings.

Wear is mechanical, which distinguishes it from chemical attack. Chemical attack and wear produce similar results and may occur together, but they remain separate problems with separate solutions. This chapter deals with wear, and Chapter 21 discusses chemical attack.

Though every floor wears, the biggest problems show up in industrial buildings (Figure 20.1). In non-industrial buildings, most floors hide under floorcoverings. While that does not eliminate wear, it effectively transfers the problem from the concrete slab to the floorcovering. Coverings can be replaced and are not expected to last for the life of the building.

In industry, however, the concrete slab usually provides the wearing surface. A structural element, meant to last for many decades, is then subject to heavy use and wear. That makes the industrial floor unique in building construction, challenging designers and builders alike.

We in the concrete-floor industry have not always met that challenge. Lack of wear resistance is a common cause of floor problems. In had cases the whole surface is destroyed to a depth of many millimetres (1/2 in or more), leaving the floor unusable. Even mild wear creates dust, which is unacceptable in some factories and warehouses.

The situation is so bad that some users won't even consider using a plain concrete floor. Believing that concrete is inherently poor at resisting wear, they spend fortunes on sealers and dust-proofers, many of which don't even work.

On the other hand, some floors resist wear very well. Many plain concrete floors last thirty years or more, even under heavy industrial traffic.

Figure 20.1 Premature wear on an industrial floor

The traditional approach

Though wear resistance is crucial, tradition floor specifications say little about it. If they address the subject at all, most designers still use method specifications rather than performance specifications.

The traditional approach is to try to control wear resistance through one or more of these contributing factors:

- concrete compressive strength;
- water–cement ratio;
- curing;
- finishing methods.

This indirect approach sometimes work, but it is risky. Many factors affect wear resistance, and no specification can control them all directly. One floor may have good concrete strength, but wear out because of poor curing. Another may have good strength and good curing, and yet wear out because the floorlayers worked bleed water into the surface.

Where resistance to wear is particularly important, the traditional response is to specify surface treatments such as toppings, coating and hardeners. Of the hundreds of products on the market, most fit into these groups:

- high-strength toppings (most often made of strong Portland-cement concrete, but sometimes of epoxy concrete or other materials;
- dry shakes (metallic or non-metallic);
- resin seals (polymer resins that coat and penetrate the floor surface);
- liquid hardeners (chemicals that are supposed to reach with concrete to harden the surface).

Surface treatments often succeed in preventing excessive wear. But there are serious drawbacks.

The main problem is price. The more effective methods are dear, some costing as much as the rest of the slab.

Another drawback is that surface treatments sometimes fail for reasons that have nothing to do with wear resistance. Toppings, dry shakes and resin seals do not always bond to the underlying concrete. If a topping peels off, the results may be worse than the wear it was meant to prevent.

There is a better way, which is to specify wear resistance and enforce it with performance tests on the finished floor.

Classifying wear resistance

British and American standards classify wear resistance differently. Neither method is wholly satisfactory, nor requires testing to see whether a floor belongs in its class. However, both methods are easily combined with the compliance test described below.

The British classification

BS 8204:Part 2 divides concrete floors into four classes, to which a fifth, called Nominal, is sometimes added. Most concrete floors fall into the three main AR (for abrasion resistant) classes, ranging from AR1 (very high wear resistance) to AR3 (good wear resistance).

The special category is for floors that need extraordinary resistance to wear. Special floor are used in heavy industry and, sometimes, in very busy warehouses.

Table 20.1 British classification of wear resistance

Class	Degree of Wear Resistance	Maximum Wear Depth*	Typical Use	Traffic
Special	Extremely high	0.05 mm 0.002 in	Very heavy-duty factories	Heavily loaded steel tyres, impact, dragged loads
AR1	Very high	0.10 mm 0.004 in	Heavy-duty factories and warehouses	Steel tyres, impact
AR2	High	0.20 mm 0.008 in	Medium-duty factories and warehouses	Lightly loaded steel tyres, hard plastic tyres
AR3	Good	0.40 mm 0.016 in	Light-duty factories and warehouses	Rubber tyres
Nominal		0.80 mm 0.031 in	Construction traffic	Pneumatic tyres, feet

*From standard test with Chaplin abrasion tester.

The Nominal category is for floor that get little wear. It is not part of BS 8204, but was in the original Cement and Concrete Association proposal that led to the standard classification. This category is for floors that will not get wear in the finished building, but still need a modest degree of wear resistance to withstand construction traffic.

Table 20.1 shows the British classification.

ACI 302 classification

This divides floors into nine categories (see Table 20.2). The classes are numbered generally in order of increasing need for wear resistance, but the list is complicated by the fact that some of the higher numbers refer to toppings, which may or may not need good resistance to wear. The highest number is for superflat toppings, which generally do not need extraordinary resistance to wear.

Though the classification comes from ACI 302, the recommended wear depths are my own. ACI does not endorse any performance standard for measuring wear resistance, even though ASTM publishes several test methods, including one, ASTM C 779, not that much different from the Chaplin test.

Table 20.2 American classification of wear resistance

Class	Use	Usual Traffic	Maximum Wear Depth*
1	Exposed concrete in offices, houses, shops and institutions	Feet	0.80 mm 0.031 in
2	Covered concrete in offices, houses, shops and institutions	Construction vehicles	0.80 mm 0.031 in
3	Toppings in non-industrial buildings	Feet	0.80 mm 0.031 in
4	Institutions and retail stores	Foot and light vehicles	0.40 mm 0.016 in
5	Medium-duty factories and warehouses	Pneumatic tyres, soft plastic tyres	0.20 mm 0.008 in
6	Heavy-duty factories and warehouses	Heavy industrial, impact	0.10 mm 0.004 in
7	Bonded toppings in heavy-duty factories and warehouses	Hard tyres and heavy loads	0.05 mm 0.002 in
8	Unbonded toppings	As for classes 4–6	0.40–0.10 mm 0.016–0.004 in
9	Superflat floors	Hard plastic tyres in defined paths	0.10 mm 0.004 in

*From standard test with Chaplin abrasion tester.

Testing wear resistance

Over the years, researchers have built dozens of machines for measuring wear. Some abraded the floor with rolling wheels. Others used abrasive pads, dressing wheels, shot-blasting or a combination of methods. In some tests, depth gauges determined the amount of wear. In others, a concrete sample was weighed before and after the test to reveal how much material was lost.

These testers were built for research. They did not become standard instruments for measuring wear resistance on site, though some of them could have been so used.

In America, several machines have been incorporated into standard tests. ASTM C 779 describes three procedures for measuring the wear resistance of "horizontal concrete surfaces". Procedure A uses rotating and revolving discs with silicon carbide grit. Procedure B uses steel dressing wheels and procedure C relies on heavily loaded ball bearings.

Other ASTM tests deal with wear resistance but are less suitable for floors. ASTM C 418 describes a method for testing small concrete samples

338 Design and Construction of Concrete Floors

by sandblasting. ASTM C 944 describes another dressing-wheel test, but one suited for small samples such as test cylinders. Two ASTM standards describe ways to test the wear resistance of coarse aggregates (not concrete) by tumbling them in a special drum called a Los Angeles machine. But as we shall see, coarse aggregate has little to do with wear resistance at the floor surface.

It is likely that one or more of these American tests could be used for or adapted to compliance testing as part of a performance specification for wear resistance. But that has not happened yet. The ASTM tests remain little used.

The Chaplin abrasion tester

In the early 1980s, the Cement and Concrete Association (now British Cement Association) developed an instrument specifically for measuring the wear resistance of concrete floors. Some early documents called it the accelerated abrasion apparatus. It was refined and is now available commercially as the Chaplin abrasion tester (Figure 20.2).

Figure 20.2 Chaplin abrasion tester

The devices spins steel wheels in a circular path to simulate the wear caused by industrial vehicles. In a 15-minute test, the wheels wear a ring-shaped groove in the floor (see Figure 20.3). A gauge measures the depth of the groove. A standard survey consists of three tests at different locations.

Depth of wear is reported in hundredths of a millimetre, with higher readings meaning lower resistance to wear. Most concrete floors shows wear depths between 0.02 and 1.00 mm.

The test has some limitations. Because it uses rolling wheels, it is best suited to floors on which wheeled vehicles are the main source of wear. Chaplin results seem to correlate reasonably with sliding wear, as well. But the Chaplin test may not be appropriate where surface damage comes from impact, corrosive wear (wear plus chemical attack) or erosion from moving liquids.

Another limitation is that the test does not produce enough wear on extremely hard floors. Some Special-class floors show wear depths of only 0.01 mm or so – right at the limit of accuracy of the depth gauge. For this reason, the standard does not discriminate well between two highly resistant floors. Two ways have been proposed to work around that limitation. One way is to lengthen the length from 15 to 30 minutes or more. Another way is to replace the smooth wheels with toothed dressing wheels.

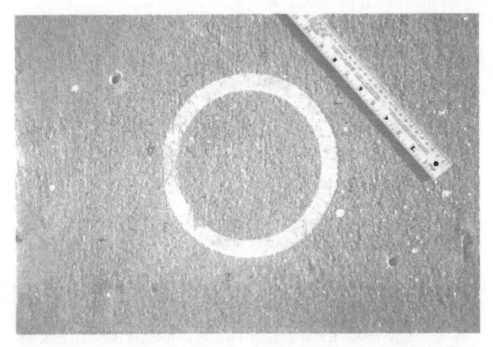

Figure 20.3 The Chaplin abrasion tester wears a ring-shaped groove in the floor

How to specify wear resistance

Writing a performance specification for wear resistance sometimes requires only two steps:

1. Choose a floor class from BS 8204 or ACI 302.
2. Specify the maximum depth of wear from Table 20.1 or 20.2, when the floor is tested with a Chaplin machine at an age of 28 days or more.

If a floor seems not to fit any standard class – and some do not – test results from a similar floor may suggest a suitable Chaplin result to specify.

The maximum wear depths are for concrete at least 28 days old. Because concrete grows stronger as it ages, wear resistance usually improves over time.

Factors that affect wear resistance

Many factors affect wear resistance. The most important seem to be finishing methods, curing, fine aggregate and certain surface treatments. Contrary to common belief, concrete strength and hardness of the coarse aggregate are less important.

We shall look at the effects of:

- concrete compressive strength;
- finishing methods;
- fine aggregate;
- coarse aggregate;
- dry shakes;
- liquid hardeners;
- resin seals;
- curing;
- carbonation.

Concrete compressive strength

Many people believe the concrete's compressive strength largely controls its resistance to wear. Some designers seem to think strength is the only factor worth worrying about.

Surprisingly, research reveals only a weak connection between concrete strength and the wear resistance of the finished floor. The connection is

Resistance to Wear

weak because wear occurs at the surface, which may be much stronger or much weaker than the bulk of the concrete below. Big differences in concrete strength – one study looked at 36 versus 85 MPa (4200 versus 9900 psi) – do show up in wear tests (Chaplin, 1986, pp. 10-11; Scripture, Benedict and Bryant, 1953, p. 316). But within the strength range found in ordinary floors, other factors – notably finishing and curing – dominate. Concrete with compressive strength ranging from 25 to 40 MPa (3000 to 5000 psi) has produced floors in every wear class from Nominal to Special.

Finishing methods

These matter more than concrete strength. Good floorlayers take justifiable pride in their ability to give concrete a durable finish by floating and trowelling.

A smoother finish means greater resistance to wear. This is true not just for concrete; it seems to be a general principle applicable to most if not all materials. All else being equal, a power-trowelled floor will resist wear better than a floated or broomed floor. And among trowelled finishes, it is the burnished finish produced by multiple passes that resists wear best. Figure 20.4 shows the marked improvement possible with multiple power trowellings.

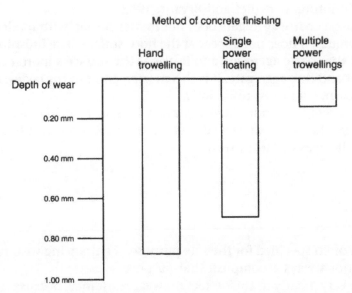

Figure 20.4 Effect of finishing methods on wear resistance

For floors that must meet wear class AR2 or better, it is risky to specify or allow anything less than a burnished finish produced by power trowelling.

Fine aggregate

The choice of fine aggregate affects a floor's resistance to wear. Chaplin (1986, p. 11) compares manufactured sand (crushed limestone fines) to several types of natural sand. When tested with a Chaplin machine, the slabs made with crushed limestone wore about three times as much as those made with natural sand.

BS 8204:Part 2 requires natural sand for AR1 and Special floors.

Coarse aggregate

In contrast to the fine aggregate, and contrary to some long-held beliefs, coarse aggregate seems to have little or no effect on wear resistance. Wear tests show no significant difference between concrete made with soft and hard coarse aggregates. Even materials as different as slag and marble on the one hand and granite and traprock on the other, give similar results. (Scripture, Benedict and Bryant, 1953)

The reason coarse aggregate does not matter is that, with modern finishing techniques, it does not appear at the floor surface. The finishing process depresses the coarse aggregate and leaves at the surface a mortar consisting of cement and fine aggregate. It is this mortar, and not the coarse aggregate farther down, that must resist wear.

Coarse aggregate undoubtedly plays a role in resisting wear after the mortar layer has been destroyed. But by then the floor will have already failed in the eyes of most users.

Dry shakes

These are often specified for the sole purpose of enhancing wear resistance. They do not always accomplish that purpose.

Shakes vary greatly in their effect on wear resistance, because they come in two different forms: non-metallic and metallic. Non-metallic shakes are brittle mineral aggregates – that is, rock – often traprock or quartz. Metallics consists of iron filings.

The non-metallics cost less but are far less effective. Research reveals little benefit from dry shakes made of brittle mineral aggregates, no matter how hard the rock. (Scripture, Benedict and Bryant, 1953; Sadegzadeh and Kettle, 1988)

In contrast, metallic dry shakes unmistakably improve a floor's resistance to wear. Test show that a metallic shake can be several times as resistant as plain concrete. (Scripture, Benedict and Bryant, 1953; Chaplin, 1986) Tested with a Chaplin machine, some metallic-shake floors show less than 0.01 mm of wear.

Such results would make a metallic shake the finish of choice for industrial floors, except for a few drawbacks. Dry shakes are costly; they sometimes delaminate; and they tend to make floors less flat.

Non-metallic dry shakes are hard to justify if their only purpose is to resist wear. (They may make sense for other purposes, such as to colour a floor.) Metallics are worth considering for floors that need an extraordinary degree of wear resistance.

There is little reason to use any dry shake on floors in wear class AR1 and below.

Liquid hardeners

These are salts, typically sodium silicate or magnesium fluorosilicate – dissolved in water. They have a small but measurable effect on wear resistance. Sadegzadeh and Kettle (1987) report that silicate hardeners improved the wear resistance of poorly cured slabs by 15–38%.

Hardeners are cheap and may be useful remedies where a modest improvement in wear resistance is called for. For example, if a floor shows wear depth just over the maximum allowed – say, 0.23 mm when the specification says 0.20 mm – then a liquid hardener just might bring the floor into compliance. You cannot expect much more than that, however.

Though some designers specify liquid hardeners in new construction, the practice is hard to justify.

Resin seals

These coat and impregnate the concrete surface with resins. The most common resins are polymers such as epoxy, polyurethane and acrylic.

Some resin seals do very well in wear tests. Sadegzadeh and Kettle (1987) report on one experiment in which a moisture-curing polyurethane resin shifted floor slabs from the Nominal class or worse to class AR2.

The drawbacks to resin seals are cost and the risk of delamination. Chapter 22 shows how to prepare a concrete surface so a resin seal is more likely to stick.

Curing

This has a huge effect on wear resistance. Chaplin (1986), working in the UK, suggests that floors cured under polyethylene can be several times more wear resistant than uncured floors (see Figure 20.5). The American Fentress (1973) reports similar results for floors cured under wet hessian (burlap).

Some resin-based curing compounds produce extremely good test results. They seem to have a double effect. Besides keeping in moisture, as does any good cure, the resins fill pores in the concrete.

Carbonation

This occurs when carbon dioxide in the air reacts chemically with fresh concrete. It can ruin a floor, resulting in a crumbling, friable surface layer that can, in the worst cases, go 12 mm (1/2 in) deep.

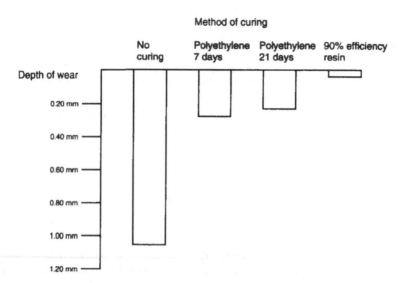

Figure 20.5 Effect of curing on wear resistance

The usual culprit is an unvented heater. Some people think that natural-gas or propane heaters are acceptable because their exhaust is (more or less) safe to breathe, but that is wrong. If a heater burns fossil fuel and vents its exhaust within the building, it can cause carbonation. Internal-combustion engines, used on concrete trucks and trowel machines, also contribute.

There is no safe level of carbon dioxide. Even the normal quantity of carbon dioxide found in the atmosphere makes concrete slightly weaker than would otherwise be the case. You can't do anything about that, but you can minimize the contribution from other sources.

Some curing methods keep out carbon dioxide and may be worth considering if no other choice exists. But the right solution is to avoid unvented heaters at all costs, and to provide ventilation for fuel-burning engines used within the building.

This harmful early carbonation differs from the long-term carbonation that occurs with hardened concrete. Early carbonation is always bad, but the long-term reaction makes concrete slightly stronger.

Improving wear resistance

There are several ways to improve the wear resistance of an existing floor. Since they do not all work on every floor, repair often comes down to trial and error. The use of a Chaplin machine or other wear tester, will help determine which methods works best.

The options for reducing wear include:

- grinding;
- late curing;
- surface treatments;
- toppings;
- cleanliness.

Grinding

Poor wear resistance often stems from a surface that is markedly weaker than the underlying concrete. This can result from poor finishing, lack of curing, early freezing or carbonation.

If only the surface is weak, grinding often helps. The process is similar to the early-age grinding used as a finishing method, described in Chapter 17. The grinding must go deep enough to remove the weak layer; 1.5–3.0 mm

(1/16–1/8 in) will usually suffice. Large-scale grinding should not start till a trial proves that grinding will in fact produce the degree of wear resistance needed. Generally you only have to grind a square metre (square yard) or two, with wear tests before and after, to find out what works.

You can sometimes save money by grinding only those parts of the floor that get traffic. In a warehouse, grinding would rarely be needed under racks.

Grinding may reduce floor flatness. Where flatness matters, you may have to survey the floor during grinding to ensure the ground-down surface meets reasonable tolerances.

Late curing

Where poor curing is the reason for a weak surface, it is sometimes possible to re-start the curing process by flooding the floor with water.

Chaplin (1986) describes a test in which slabs were air-dried for 28 days, and then soaked and cured under polyethylene for 7 days. As shown in Figure 20.6, the late curing substantially improved the slabs' resistance to wear. However, not even the late-cured concrete was as resistant as concrete-cured well from the start.

Surface treatment

We have already considered the effect of liquid hardeners and resin seals on new floors. Both kinds of treatments are also used to repair existing floors.

Liquid hardeners are relatively cheap and easy to use, but produce at best only a modest improvement in resistance to wear. Hardeners are most effective on concrete with a low water–cement ratio, a condition often not met on floors with poor wear resistance.

Resin seals work better, but usually cost more and can fail by delaminating unless the concrete is well prepared. Good preparation is sometimes hard to accomplish on existing floors.

Toppings

A concrete topping (see Chapter 18) can give almost any desired degree of wear resistance. This should be the method of last resort, however.

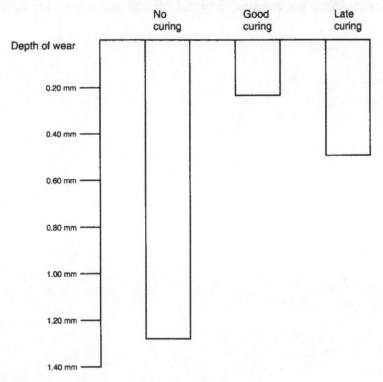

Figure 20.6 Effect of late curing on wear resistance

Toppings are dear and take days to lay. In bonded topping, there is always some risk the bond will fail. An unbonded topping eliminates that risk, but raises the finished floor elevation by at least 100 mm (4 in).

Cleanliness

Clean floors last longer than dirty ones. Loose debris leads to what is called three-body wear, in which the debris particles roll between the traffic and the floor surface. Depending on what the debris consists of, this kind of wear can be severe. And some kinds of debris lodge in vehicle tyres, greatly increasing wear.

None of this excuses bad design or bad construction practices, of course. We need to design and lay floor that have the wear resistance they need for the intended use. Still, when a user faces a rapidly wearing floor, a regular cleaning programme may put off the day repairs are needed.

21

Resistance to Chemical Attack

Most floors do not suffer greatly from chemical attack. But some do – mainly but not exclusively in factories. The attacks range from occasional spills to near-constant exposure.

Concrete's chemical resistance

Plain concrete resists alkalis and mineral oils very well. But the following chemicals damage it with ease:

- acids;
- vegetable oils and fats;
- sugar solutions.

All three groups occur often in food processing plants. Acids are used in many other industries as well. Even warehouse floors sometimes come under acid from acid, because of the lead-acid batteries that power many lift trucks.

We cannot do much to make plain concrete more resistant to chemicals, because we are dealing with basic properties of the cement and aggregates (especially the cement). We are not completely helpless, however. These steps all help concrete resist chemical attack:

- higher cement content;
- lower water–cement ratio;

- good compaction;
- good curing;
- the smoothest possible finish.

They all work by making the concrete less porous. Dense concrete is less pervious to liquids, and thus better able to resist liquid chemicals.

Another option is to make the concrete out of something beside hydraulic cement. Epoxy or sulphur can serve as cement, producing a concrete that resists some chemicals much better than ordinary concrete made with Portland cement.

Epoxy concrete is never used in full-thickness slabs because it costs too much and generates too much heat as it cures. It works well as a topping, however, making a hard-wearing floor with excellent resistance to most acids and other chemicals. Due its extreme stickiness before it has hardened, epoxy concrete is hard to finish. Some contractors let it harden and then finish it by sanding, followed by application of a final thin coating.

Sulphur concrete, in contrast, can replace Portland-cement concrete throughout the slab. It is made by mixing dry aggregates with elemental sulphur. The batch is then heated to sulphur's melting point and placed in the forms. It hardens by cooling, and can be worked with normal concrete tools. The finished product, with structural properties similar to those of ordinary concrete, resists most acids and a wide variety of salts. It has low resistance to alkalis, however.

Protecting the concrete

Since we can do so little to improve concrete's inherent resistance to chemical attack (unless we adopt exotic cements such as sulphur and epoxy), the main defence must be to minimize contact between the concrete floor and aggressive chemicals.

Toppings and coatings

A very common method of protection is to cover the concrete slab with a chemical-resistant topping or coating (Figure 21.1). Manufacturers sell a wide range of products for that purpose, including:

- polyurethane paints;
- epoxy paints;

Figure 21.1 An epoxy coating resists many forms of chemical attack

- epoxy-mortar toppings;
- epoxy-concrete toppings;
- polyester-mortar toppings;
- latex-modified concrete toppings;
- asphalt toppings.

No product works for every application. Few standards apply here. We have little choice but to read the manufacturers' brochures and decide on one proprietary products or another. Newcomers to this specialized field should proceed with caution, because it is easy to spend a huge amount of money on a coating or topping that does not work.

Tiles

In the food processing and some other industries, special tiles or bricks are often laid over the concrete floor. They can work very well, but the covering must be continuous. Even one open joint or crack can let damaging chemicals reach the underlying concrete, where they can work undetected till the slab is ruined.

The grout between tiles is often the weakest link. Because ordinary tile grout contains Portland cement, it doesn't resist chemicals any better than does ordinary concrete. Grouts made with epoxy or furan resin work better in aggressive environments, but can be tricky to install. (Aalberts, Baty and Larkin, 1989) Furan grout comes only in black.

Sealing joints and cracks

It does no good to coat a concrete floor with highly resistant material if joints and cracks remain open gateways for aggressive liquids. There have been some interesting failures in which liquids got past an epoxy topping and ate away the concrete underneath, leaving almost nothing to support the topping.

One solution is to fill all gaps, and then cover everything with a chemical-resistant coating. But that does not always work, because materials tough enough to serve as a wearing surface often lack the elasticity to stay intact as joints and cracks widen over time.

The alternative solution is to carry joints and cracks through the coating or topping and then fill them, choosing a chemical-resistant caulk. The joints and cracks then remain visible, though filled, at the floor surface.

Cleanliness

The degree of chemical attack is closely tied to how long the chemical stays in contact with the floor. Even highly aggressive spills do little harm if they are cleaned up right away.

When deciding whether a concrete floor needs a coating, topping or floorcovering to protect if from chemical attack, we should ask these questions:

- Will spillage be rare, occasional, continual?
- Will spills be cleaned up promptly?

If spillage will be continual, the floor probably needs special protection. But if spillage be rare or occasional, it may make sense to use a plain concrete floor and provide for cleaning-up. Some floors in harsh environments are sloped to drains. When the floor users detect a spill, they immediately sluice it down with water, minimizing the time the attacking chemical can act on the concrete floor.

Attack from below

So far we have looked only at attack caused by chemicals spilled from above. Danger can also come from below, caused by chemicals dissolved in ground water.

The main risk here comes from sulphate-bearing ground water. We have at least three defences against it:

- good, strong concrete;
- sulphate-resisting cement;
- a damp-proof membrane under the slab.

Sound, strong, well-compacted concrete can go a long way toward resisting sulphate, even when the mix contains ordinary Portland cement. According to one authority, a concrete mix designed to resist sulphates should have a cement content of at least $280\,kg/m^3$ ($470\,lb/yd^3$) and a water–cement ratio not greater than 0.55. (Blackledge, 1975) Most concrete used in floors meets both requirements anyway.

For more protection, consider replacing ordinary Portland cement (American Type I) with sulphate-resisting cement (American Type V), if you can find it. Americans have the added option of using Type II cement, which resists sulphates better than Type I, though not quite so well as Type V. Type II cement is readily available and is in fact the normal, everyday cement in some parts of the USA.

A good damp-proof membrane keeps sulphate-bearing ground water from reaching the concrete slab. The usual membrane – thin plastic sheeting – does not suffice for all conditions, however. In severe environments a rubber membrane with sealed joints provides better protection.

22

Preparation for Coatings, Toppings and Floorcoverings

A theme of this book is that plain, direct-finished concrete slabs serve well for many purposes. Nevertheless, many floors need to be covered. In houses and offices, floorcoverings are almost universal. In warehouses and factories, some floors get coatings or toppings to give them properties hard or impossible to achieve with plain concrete.

Though some floorcoverings can be put down on almost any floor, others have strict requirements. Here are the main points to consider when a concrete slab will receive a coating, topping or floorcovering:

- moisture;
- bond;
- surface regularity.

Moisture

Water lies at the root of most floorcovering failures. Damp concrete under floorcoverings leads to rot, unsightly mildew and growth of dangerous moulds. Moisture makes toppings and coatings peel off, and wrecks the glue under vinyl tile and stuck-down carpet.

Reading some modern publications on floor design and construction, you might get the idea that protecting floors from moisture is not very important. Do not make that mistake. This is a huge problem. The reason some writers say little about it is that they focus on the direct-finished, bare concrete floors used in warehouses and factories. Moisture does not harm those floors. But if you deal with residential, office or commercial floors – or indeed with any slabs that receive floorcoverings – you need to think about moisture.

Apart from floods and roof leaks, which lie beyond the scope of this book, damaging moisture comes from two sources: the groundwater under the floor and the mixing water added to the concrete during construction.

Moisture from below

The ground below the floor always contains some water. In moist climates it can contain quite a lot of water, reaching the point of total water saturation – called the water table – just a metre or two below the surface.

If you lay a concrete floor directly on the ground, water vapour rising through the ground will tend to accumulate on the underside of the slab. But it will not stop there, because concrete is somewhat pervious to water vapour. Some moisture will pass through.

What happens next depends on what lies on top of the concrete floor. If the concrete is bare, water vapour will pass into the air and never be noticed. But if the concrete is covered with a material less pervious to moisture, water will collect under the covering. Or if the concrete is covered with a moisture-sensitive material, water will enter that material. Then damage ensues.

To prevent that damage, some floors need protection from rising water vapour. Table 22.1 classifies floorcoverings according to their vulnerability. Coverings with no vulnerability can be laid with little or no concern for protection. Materials having "moderate" vulnerability are sometimes laid without protection, especially in drier regions. Coverings with high vulnerability should always get protection from damp.

Concrete itself – including concrete toppings – is not damaged by moisture. Bare concrete slabs normally need no protection from damp, except where the ground water contains sulphates. Rising water vapour has caused debonding of concrete toppings, however, where a latex bonding agent was used between base slab and topping.

Damp-proof membranes

Damp-proof membranes provide our main defence against water vapour rising from the ground. Americans call them vapour barriers or vapour retarders.*

---

*Vapour barrier is the traditional term, but it offends some people who argue for "vapour retarder" because the materials normally used do not totally bar the transmission of water vapour. They just slow it down. Well, okay, but I still say "vapour barrier" on the grounds that few will misunderstand me and that the English language regularly uses "barrier" to describe things that impede but do not totally prevent. What about the Great Barrier Reef?

Table 22.1 Vulnerability of floorcoverings to moisture damage

Floorcovering	Vulnerability		
	None	Moderate	High
Access flooring	X		
Anti-static vinyl			X
Asphalt topping	X		
Bitumen topping	X		
Brick	X		
Carpet			X
Cement–sand screed	X		X
Ceramic tile	X		
Composite vinyl tile		X	
Concrete topping	X		
Cork			X
Epoxy coating		X	
Epoxy topping		X	
Felt-backed vinyl			X
Hardwood			X
Linoleum			X
Magnesite (magnesium oxychloride)			X
Polyester resin			X
Polyurethane coating		X	
Rubber sheet			X
Semi-flexible vinyl tile		X	
Stone	X		
Terrazzo	X		
Vinyl sheet			X
Vinyl tile		X	
Wood block (decorative, laid in cold glue)			X
Wood block (industrial, laid in hot mastic)	X		
Wood strip or plank			X

Note: Invulnerable materials may become vulnerable if stuck down with moisture-sensitive glues.

The common material for this purpose is polyethylene sheet. The UK's Building Research Establishment (1979) recommends polyethylene at least 0.12 mm (5 mils) thick. Americans use a similar thickness – generally 0.10 mm (4 mils) or 0.15 mm (6 mils). Some designers specify thicker polyethylene, partly for its lower permeability, but mainly for its ability to withstand construction traffic without damage.

Polyethylene has come under attack in recent years. Critics argue that it lets too much water vapour pass through, that it rips too easily during construction, and that it deteriorates underground. No doubt there is some truth behind those criticisms. Nevertheless, polyethylene remains the damp-proof membrane of choice, and it seems to work adequately in most floors, most of the time.

Where designers want something better than polyethylene sheet, they normally specify a high-performance, proprietary membrane with sealed seams. Some of those membranes resemble the materials used to waterproof flat roofs. They all cost far more than polyethylene sheet.

Before switching from polyethylene sheet to any other material, make sure the floor design does not rely on the polyethylene to do double duty as a slipsheet, reducing friction between slab and ground. Some alternative membranes actually increase friction.

We can specify damp-proof membranes by performance, though few designers do. ACI 302 recommends that damp-proof membranes have permeability no greater than 0.2 perm (0.3 US perm) when tested according to ASTM E 96.

The damp-proof membrane may be installed at any of the following locations:

- directly under the concrete slab;
- between the sub-base and the fines layer;
- under the sub-base;
- between the slab and an unbonded topping or screed;
- directly under the floorcovering.

Each location has benefits and drawbacks. The most common location is right under the concrete slab. This is the only location where the membrane can reduce friction between slab and ground – an important factor in some designs. Drawbacks include increased bleeding and an increased risk of slab curl.

To reduce bleeding and curling, we can put the membrane lower and cover it with a thin layer of fines. If dry, the fines act as a blotter, soaking up excess water from the bottom of the slab. This approach is risky on outdoor pours, however. If rain falls after the fines are in place, water becomes trapped above the membrane. Another drawback is the risk that fines may conceal rips and holes in the membrane caused by construction traffic and setting forms.

Some designers go even further, putting the membrane under the sub-base. That reduces the risk of concealed damage to the membrane, since a thick sub-base provides good protection. It is still vulnerable to rain.

On those few floors that get an unbonded topping (or unbonded cement–sand screed), a good option is to lay the membrane between the slab and the topping. That not only blocks water vapour, but improves the performance of the topping by isolating it from the slab and reducing friction. New floors rarely get unbonded toppings, but this can be a solution where an existing floor lacks a damp-proof membrane and needs one.

The last choice is to put the membrane between the concrete floor and the floorcovering. That would not work with most coverings, but is a reasonable option for wood flooring laid on sleepers, and for some other kinds of flooring that leave an air gap between the slab and the finish floor.

Alternatives to damp-proof membranes

A damp-proof membrane is not the only way to protect a floorcovering from ground moisture.

Some American designers (by no means all) have strong feelings against membranes. They omit the vapour barrier in situations where British designers would leave it in. Instead of a membrane, they make sure the site is well drained and use free-draining coarse aggregate under the slab to create a capillary break.

Another possible alternative is to add a waterproofing chemical to the concrete mix, but the Building Research Institute (1979) recommends against it.

I would not presume to tell designers exactly which floors need membranes, but I will say this much: the consequences of omitting a membrane where it is needed are far worse than the consequences of using one where it is not needed.

What to do when the membrane is missing

Sometimes you need to lay a floorcovering where a damp-proof membrane should have been, but was not, installed. The membrane may have been forgotten, or the floor usage may be changing from one that did not need protection from water vapour. It is not a good situation, but neither is it hopeless.

Consider these remedies:

- Cover the slab with an unbonded topping or unbonded cement–sand screed, using a membrane between slab and topping.
- Install a damp-proof membrane between slab and floorcovering. This only works for a few kinds of coverings, such as wood flooring laid on sleepers.

- Choose a floorcovering not easily damaged by moisture. Some choices, such as ceramic tile, brick and stone, are immune to water damage.
- Test the concrete for moisture content to see if it really is damp enough to cause trouble.

If you try the last suggestion and test for moisture, use judgment when interpreting the test results. Readings change over the course of a year and from year to year. A good test result taken near the end of a five-year drought would not prevent problems the following year when the site gets record rainfall.

Moisture from within

Water vapour arising from the ground is far from the only source of damaging moisture. Even on the driest sites, every concrete slab, concrete topping or cement–sand screed starts life soaking wet. The excess water eventually evaporates, but till it does it can interfere with moisture-sensitive floorcoverings.

The key is to let the concrete dry out before installing a floorcovering, but that can take a long time. One rule of thumb says we should wait four weeks for every 25 mm (1 in) of slab or screed thickness. Following that rule, we would have to wait 16 weeks before covering the typical 100 mm (4 in) house or office slab. Thick industrial slabs would need even more time.

That four-weeks-per-25-mm rule is just a rough guide, however. Slabs dry out much faster in hot, dry weather and in winter when the building is heated. On the other hand, the process takes much longer in cold, damp conditions and on outdoor pours where the floor is repeatedly drenched with rain.

Instead of waiting a prescribed length of time, some floorcovering contractors prefer to test the concrete's moisture content. A simple but reasonably effective test is to tape down a small square of polyethylene sheet overnight. If you see liquid water the next day, the concrete is probably too wet. Some contractors get instant results using hand-held testers that measure the concrete's electrical resistance, giving a simple "pass" or "fail".

The British Cement Association suggests testing with an hygrometer, which measures air's relative humidity, as follows:

1. Place an hygrometer in a vapour-tight box open at the bottom.
2. Set the box on the floor, open side down and seal it to the floor with tape.
3. Measure the relative humidity within the box.

Preparation for Coatings, Toppings and Floorcoverings

The Association says a reading of 80% is generally safe, but recommends 75% or less to allow for measurement error.

Bond

Most floorcoverings are glued or stuck down in some way to the concrete. Some coverings, such as carpet and vinyl tile, will stick well enough to any concrete surface that is clean and dry. But coverings in the following list require an extremely good bond:

- bonded concrete toppings (see Chapter 18);
- bonded cement–sand screeds (see Chapter 18);
- epoxy-concrete toppings;
- polymer-resin coatings, including epoxies and polyurethanes.

To bond well to those coverings, the concrete surface should be at least slightly rough, free of weak laitance and very clean. Contractors use the following methods to prepare concrete for bonding:

- scabbling;
- shot-blasting;
- grinding;
- scarification;
- scratch finish;
- acid etching.

Scabbling roughens the concrete by hammering. The typical scabbler has several hardened-steel head the move up and down, driven by compressed air. Scabbling should continue till it exposes the coarse aggregate. It does a good job, but makes a lot of noise and leaves a mess. The action is too violent for some suspended floors. And the resulting surface is too rough for most coatings and for very thin toppings.

Shot-blasting erodes the concrete by driving tiny steel balls (shot) against it. The shot is recirculated and the concrete dust extracted by vacuum. Shot-blasting is clean and quiet, making it a good choice for building in use.

Grinding is not very common but works well, leaving the surface smoother most other methods, though still rough enough for a good bond. Do not use a high-speed terrazzo grinder that polishes the floor.

A *scarifier* roughens the surface with a rotating cylinder. Special washers that resemble small gears are mounted loosely on the cylinder's surface, where they bounce and rattle against the concrete, breaking off tiny chips.

Some builders prepare a floor by giving it a *scratch finish* (see Chapter 17) or by leaving a rough-tamped surface. Both methods are less reliable than scabbling, shot-blasting, grinding or scarification, all of which remove the concrete surface and can be performed immediately before the floorcovering goes on. The American Concrete Institute suggests a scratch finish may be good enough, but warns it must be kept clean or cleaned later with detergent and acid. UK authorities are even less encouraging. Deacon (1976) warns us: "Preparation is simplified if the base is left smooth and not ridged with tamp marks. *A heavily tamped surface alone does not provide an adequate key*".

In *acid etching,* the floor is scrubbed with muriatic (a chemist would call it hydrochloric) acid. This is useful for cleaning, but cannot really replace mechanical methods.

No matter how the surface is prepared, the effort will go to waste if the concrete is not clean when the covering is laid. Cleaning the concrete thoroughly is not good enough. It must stay clean or be cleaned again. Many failures have occurred because a well-prepared base slab was fouled by other construction trades. A little dust or even the overspray from nearby painting can prevent bond.

The ensure cleanliness, it is wide to install the next course as soon as possible after the base slab is prepared. Even one day may be too long to wait on a dusty building site. Some contractors cover the prepared surface with polyethylene sheet or building paper to keep it clean.

Surface regularity

This matters where the floorcovering is a thin sheet material such as vinyl tile. Minor irregularities show through, and can become even more obvious than on bare concrete. The problem also occurs with some floor coatings, especially those that are opaque and glossy. The objectionable defects fall into two categories.

The first category consists of bumps and dips having a wavelength of 0.3 to 1.2 m (1 to 4 ft). A standard flatness test will detect such defects, and the right flatness specification, if enforced, will prevent them. Unfortunately, it is hard to determine the right flatness specification, since we are dealing here with aesthetics and personal opinion. Most users are happy with an $F_F 30$ surface (between FM 1 and FM 2 in the British TR 34 system). Many live with worse. But some particularly demanding users have complained even where the surface measured above $F_F 50$ (well above FM 1).

Preparation for Coatings, Toppings and Floorcoverings 363

The second category consists of very short, sharp defects such as trowel marks, faulted joints and cracks, and poorly finished construction joints. Standard flatness tests usually miss such defects. The human eye is the only available test device.

If detected in time (before the floorcovering is laid, that is), surface irregularities can be corrected by:

- grinding;
- covering the slab with a cement–sand screed;
- application of a proprietary smoothing compound (often but inaccurately called a levelling compound);
- use of a more forgiving floorcovering.

References

This list does not include published standards identified by the letters BS, ACI, or ASTM. Text references to such standards include each standard's number.

Aalami, B. O. and Jurgen, J. D. (2003). Guidelines for the design of post-tensioned floors. *Concrete International*, March.

Aalberts, J., Baty, D. and Larkin, M. C. (1989). Furan grout and epoxy secure heavy-duty brick floor. *Plant Services*, September, 83–94.

American Concrete Institute (1982). *Concrete Craftsman Series – Slabs on Grade*. American Concrete Institute.

Blackledge, G. F. (1975). *Cements*. Cement and Concrete Association, 3.

Blackledge, G. F. (1977). *Formwork*. Cement and Concrete Association, 13.

Blackledge, G. F. (1980). *Concrete Admixtures*. Cement and Concrete Association.

Blackledge, G. F. (1980)B. *Placing and Compacting Concrete*. Cement and Concrete Association.

Blackledge, G. F. (1980). *Ready-mixed Concrete*. Cement and Concrete Association.

Branson, D. E. (1977). *Deformations of Concrete Structures*. McGraw-Hill.

British Cement Association (2000). *New eurocements: Information sheet no. 1*. British Cement Association.

Building Research Establishment (1979). *Damp-Proofing Solid Floors* (Digest 54). Building Research Establishment.

Cement and Concrete Association of Australia (2003). *Guide to Residential Floors*. Cement and Concrete Association of Australia.

Carino, N. J. and Lew, H. S. (1982). Re-examination of the relation between splitting tensile and compressive strength of normal weight concrete. *ACI Journal*, May–June, 217.

Chandler, J. W. E. (1982). *Design of Floors on Ground*. Cement and Concrete Association.

Chandler, J. W. E. and Neal, F. R. (1988). *The Design of Ground-supported Concrete Industrial Floor Slabs*. British Cement Association.

Chaplin, R. G. (1986). *The Influence of Cement Replacement Materials, Fine Aggregates and Curing on the Abrasion Resistance of Concrete Floor Slabs*. Cement and Concrete Association.

Chaplin, R. G. (1990). *The Influence of GGBS and PFA Additions and Other Factors on the Abrasion Resistance of Industrial Concrete Floors*. British Cement Association.

Concrete Construction (1989). Monitoring concrete consolidation in pavements. *Concrete Construction*, June, 592–593.
Concrete Society (2003). *Concrete Industrial Ground Floors* (Technical Report 34, Third Edition). Concrete Society.
Concrete Society (1987). *Concrete Industrial Ground Floors* (Technical Report 34), Concrete Society.
Corps of Engineers (1984). *Engineering Design: Rigid Pavements for Roads, Streets, Walks and Open Areas*. Department of the Army.
Deacon, R. C. (1986). *Concrete Ground Floors: Their Design, Construction and Finish*. Cement and Concrete Association.
Deacon, R. C. (1976). *High-Strength Concrete Toppings for Floors, Including Granolithic*. Cement and Concrete Association.
Dennis, R. (1989). Bonding screeds and topping to the concrete base. From notes to BCA course *Concrete Floors – Design and Construction to New Standards*, 15–16 February.
Department of the Environment (1988). *Design of Normal Concrete Mixes*. HMSO.
Dewar, J. D. (1986). Ready-mixed concrete mix design. *Municipal Engineering*, 3 Feb., 35–43.
Face, A. (1987). Specifying floor flatness and levelness: the F-number system. *Construction Specifier*, April.
Federal Highway Administration (1990). *Technical Advisory: Continuously Reinforced Concrete Pavement*. US Department of Transportation.
Fentress, B. (1973). Slab construction practices compared by wear test. *ACI Journal*, July.
Hayward, D. (1985). Slab firm goes for mix not strength. *New Civil Engineer*, 5 September.
Japan Society of Civil Engineers, 1985. *Method of Test for flexural strength and flexural toughness of SFRC*. Japan Society of Civil Engineers.
John Kelly Lasers (1989). Technical report 89/2/01. John Kelly Lasers.
Kass, J. L. and Campbell-Allen, D. (1973). Precision of early shrinkage measurements. *Journal of Testing and Evaluation*, January, 24–30.
Kelley, E. F. (1939). Application of the results of research to the structural design of concrete pavements. *ACI Journal*, June, 437–464.
Lotha, R. P., Nautiyal, B. D. and Jain, O. P. (1976). Creep of fly ash concrete. *ACI Journal*, August, 471–472.
Matthew, P. W. (1990). *Economic Long-Span Concrete Floors*. British Cement Association.
Metzger, S. N. (1983). How to prevent failures of industrial floors. *Plant Engineering*, 22 Dec., 57.
Miltenberger, M. A. and Attiogbe, E. K. (2002). Shrinkage-Based Analysis for Control-Joint Spacing in Slabs-on-Ground. *ACI Structural Journal*, May–June 2002, 352–359.
Murdock, L. J., Brook, K. M. and Dewar, J. D. (1991). *Concrete Materials and Practice*. Edward Arnold.
Neal, F. (1987). Industrial ground slabs. *Estates Gazette*, 19 December.
Neal, F. R. and Judge, C. J. (1987). *Classes of Imposed Floor Loads for Warehouses*. Building Research Establishment.
Phelan, W. (1989). Floors that pass the test: How to build floors that meet F-number floor flatness and levelness specifications. *Concrete Construction*, January.
Pickett, G. A study of stresses on the corner region of pavement slabs under larger corner loads. In *Concrete Pavement Design* (Portland Cement Association), 77–87.
Post-Tensioning Institute (1980). *Design and Construction of Post-Tensioned Slab on Ground*. Post-Tensioning Institute.
Raphael, J. M. (1984). Tensile strength of concrete. *ACI Journal*, March–April, 163.
Ringo, B. C. and Anderson, R. B. (1992). *Designing Floor Slabs on Grade*. Aberdeen Group.
Roberts, B. F. (1986). *Testing Cement-Sand Screeds Using the BRE Screed Tester*, Cement and Concrete Association.

Sadegzadeh, M. and Kettle, R. J. (1987). Abrasion resistance of polymer-impregnated concrete. *Concrete*, May, 32–34.

Sadegzadeh, M. and Kettle, R. J. (1988). Abrasion resistance of surface-treated concrete. *Cement, Concrete, and Aggregates*, Summer.

Salinas, J. J. (1980). Some economic implications in reinforced concrete slab design. In *Advances in Concrete Slab Technology*. Pergamon Press, 323–330.

Schrader, E. K. (1981). Impact resistance and test procedure for concrete. *ACI Journal*, March–April, 141.

Scripture, E. W., Benedict, S. W. and Bryant, D. E. (1953). Floor aggregates. *ACI Journal*, December.

Spears, R. and Panarese, W. (1990). *Concrete Floors on Ground*. Portland Cement Association.

Taylor, P. J. and Heinman, J. L. (1977). Long-term deflection of reinforced concrete slat slabs and plates. *ACI Journal*, November, 556–561.

TR 34. See Concrete Society. (2003).

Wells, J. A., Stark, R. D. and Polyzois, D. (1999). Getting better bond in concrete overlays. *Concrete International*, March.

Westergaard, H. M. (1925) Computation of stresses in concrete roads. *Proceedings of the 5th Annual Meeting of the Highway Research Board*, Part I, 90–112.

Westergaard, H. M. (1947). New formulae for stress in concrete pavements and airfields. *Proceedings of the American Society of Civil Engineers*, May, 687–701.

Wire Reinforcement Institute (1973). *Design Procedures for Industrial Slabs*. Wire Reinforcement Institute.

Glossary

accelerator – admixture that increases the rate at which fresh concrete sets and gains strength.

addition – material added to dry cement.

admixture – material added to a concrete mix, but not including cement, pozzolans, aggregates, water and fibres.

air entrainment – process by which very small bubbles are created within concrete, mainly to increase frost resistance.

aggregate – component of concrete consisting of rock particles.

aggregate interlock – method by which loads are transferred across cracks (including the cracks that form beneath induced joints) through direct contact between pieces of aggregate.

alkali-aggregate reaction – damaging chemical reaction between the alkalis in cement and certain aggregates, resulting in popouts.

armoured joint – joint fitted with steel bars or steel angles to protect the concrete edges.

bar joist – floor-supporting beam made of steel bars welded into a truss.

bearing wall – vertical panel, typically made of concrete or masonry, that forms the structural support for a suspended floor.

bituminous concrete – concrete in which bitumen, a tar-like substance produced from petroleum, serves as cement.

bleeding – segregation of water from fresh concrete. Concrete bleeds because water is lighter than most other components. Water rises to the top as other components sink.

block stacking – warehouse storage system in which unit loads, such as palletized goods, are stacked directly on the floor, with access aisles for vehicles.

blockout – formed gap around a building column or other slab penetration.

blotter layer – course of material placed beneath a floor slab and above the damp-proof membrane for the purpose of absorbing excess water from the fresh concrete.

bonded topping – layer of concrete stuck down to a hardened base slab. Also called deferred topping.

bond strength – tensile strength (resistance to being pulled apart) between concrete and another material, or between two courses of concrete.

broom finish – rough floor finish achieved by passing a broom or brush over the surface.

bull float – long-handled float designed so the user can stand off to one side of the slab. Also called skip float.

burnished finish – the smoothest possible floor finish, normally achieved through repeated passes of a power-trowel machine.

carbonation – chemical reaction between atmospheric carbon dioxide and the lime found in concrete.

cement – component of concrete that binds the aggregates into a solid mass.

cement replacement material – powder that has cementitious properties when used with Portland cement. Examples include fly ash, slag and certain volcanic ashes called pozzolanas.

chair – device for supporting reinforcement within a slab.

characteristic strength – strength for design purposes, based on the mean measured strength less an allowance for variability.

check rod – lightweight highway straightedge (q.v.) used to check floor flatness whilst concrete is soft.

coarse aggregate – aggregate consisting almost wholly of rock particles at least 5 mm (3/16 in) across.

combination blades – blades fitted to a power-trowel machine and usable both for floating and trowelling. A combination blade is rounded on its leading edge and sharp on its trailing edge.

compaction – process by which voids are eliminated from fresh concrete.

compressive strength – resistance to crushing.

construction joint – floor joint between concrete pours. Also called day joint or formed joint.

contraction joint – floor joint designed to let concrete contract or shrink.

control joint – floor joint designed to control cracks. The term is sometimes reserved for induced joints, in contrast with the construction joints between pours.

crazing – network of very fine, closely spaced cracks in a floor surface.

crusher run – the whole, unsieved output of a rock crusher, containing both coarse and fine materials. Seldom suitable for use as concrete aggregate, but often used in sub-bases.

cube strength – concrete compressive strength based on testing of cubes. This is the normal method in the UK.

curing – process by which newly placed concrete is kept moist so it can keep gaining strength.

curing compound – liquid sprayed or rolled on a floor surface to cure the concrete by reducing evaporation of water.

cylinder strength – concrete compressive strength based on testing of cylinders. This is the normal method in America and continental Europe.

damp-proof membrane – sheet material placed under a ground-level floor to reduce the amount of water vapour rising from the ground. Also called vapour barrier or vapour retarder.

darby – long float with the handle mounted near one end, used mainly at slab edges.

delamination – 1. separation of the trowelled floor surface from the main part of the slab. 2. failure of bond between a topping and its base slab.

deferred topping – layer of concrete stuck down to a hardened base slab. Also called bonded topping.

dowel – smooth bar, normally made of steel, used to transfer loads across a floor joint. Unlike tie bars, dowels let joints open as the concrete shrinks.

dowel basket – wire frame for supporting dowels at induced joints.

dry shake – mixture of dry aggregate and cement sprinkled on fresh concrete and worked in with floats. Also called sprinkle finish.

drying-shrinkage crack – crack caused by concrete's drying out after it has hardened.

ductility – ability of concrete to resist bending after it has cracked.

early-entry saw – power saw designed for use on just-hardened concrete, in contrast with ordinary concrete saws that cannot be used till later. Also called Soff-Cut saw, a trade name.

entrained air – very small bubbles created within concrete, mainly to increase frost resistance.

entrapped air – air inadvertently introduced within a concrete mix and not eliminated by compaction. Distinct from entrained air, which is deliberately introduced.

epoxy – polymer resin with many uses in concrete floors. Epoxy is used to coat floors, fill joints, repair cracks and bond toppings. It serves as the cement in epoxy concrete and epoxy mortar.

epoxy concrete – concrete in which epoxy resin replaces Portland cement as the binder.

expansion joint – floor joint that contains a gap, usually filled with compressible material, to let the concrete expand.

expansive cement – cement that expands soon after being mixed with water. Used in shrinkage-compensating concrete.

fibres – wires or threads made of steel or plastic, added to a concrete mix for reinforcement.

fine aggregate – aggregate consisting almost wholly of rock particles less than 5 mm (3/16 in) across.

finishability – ease with which fresh concrete can be straightedged, floated and trowelled.

flat plate – two-way suspended floor in which the top and bottom surfaces are parallel planes.

flat slab – two-way suspended floor with thickened sections, called drop panels, over the supports.

flatness – degree to which a floor surface approaches a perfect plane.

flexural strength – resistance to bending.

float – finishing tool with a thick, flat blade, used to smooth the concrete surface and imbed aggregates.

float finish – intermediate floor finish (smoother than a broom finish and rougher than a burnished finish) requiring that the surface be floated but not trowelled.

float shoes – wide blades fitted to a power-trowel machine, kept level with the floor surface, and used similarly to a traditional hand float.

fly ash – airborne residue from the burning of coal, used as a cement replacement material. Also called pulverized fuel ash or pfa.

form – mould, or part of the mould, into which concrete is poured.

fresno – trowel fitted to a long handle.

gap-graded – having gaps in the gradation of the coarse aggregate.

girder – beam that supports joints, which in turn supports a floor.

grade beam – concrete beam under a ground-level slab.

ground-supported floor – floor that rests on, and is supported by, the ground. Also called slab on grade.

high-range water reducer – admixture that greatly increases concrete's workability. Also called superplasticizer.

highway straightedge – tool used for flattening concrete, consisting of an extruded metal beam attached to a long handle. Also called scraping straightedge or bump cutter.

hydraulic cement – powder that forms a strong, solid mass when mixed with water. Portland cement is the most common type.

induced joint – floor joint created within a concrete pour, usually by sawing partway through the slab thickness or by casting an insert within the pour.

integral topping – layer of concrete placed over a just-placed base slab. Also called monolithic topping.

internal vibrator – tool for compacting concrete by vibration and meant to be inserted into the wet concrete. Also called immersion vibrator or poker vibrator.

isolation joint – detail separating a concrete floor from another building element such as a wall or column.

joint – seam within a concrete floor, or between a concrete floor and another building element.

joint filler – 1. hard material installed in floor joints to protect the concrete edges from damage. 2. compressible material installed in isolation and expansion joints.

joint sealant – caulk installed in a floor joint to keep out liquids and debris.

joist – beam that directly supports a suspended floor.

Laser Screed (trade name) – mechanized tool for striking off floors using a horizontal screw kept level by reference to a laser beam.

jointer – hand tool for making induced joints.

levelness – degree to which a floor surface is parallel to the horizon.

liquid hardener – one or more salts dissolved in water, applied to hardened concrete to increase wear resistance. Also called surface hardener.

mat foundation – ground-supported floor slab designed to support the building.

metal deck – corrugated steel sheet used to support and reinforce a suspended concrete floor. Also called profiled steel decking and (colloquially) wrinkled tin.

microsilica – very fine powder, a by-product of making silicon alloys, used as a pozzolan. Also called silica fume.

moist curing – curing method in which concrete is kept constantly moist.

monolithic topping – layer of concrete placed over a just-placed base slab. Also called integral topping.

one-way floor – suspended floor designed to span in one direction.

overslab – layer of concrete placed over but not bonded to a base slab. Also called unbonded topping.

pan float – metal disc fitted to a power-trowel machine and used for floating.

pipe screed – pipe supported at finish floor elevation and used as a temporary guide for striking off.

plastic-settlement crack – crack in floor surface caused when plastic concrete sinks relative to a fixed object such as a reinforcing bar.

plastic-shrinkage crack – crack in floor surface caused when plastic (that is, wet) concrete shrinks.

plasticizer – admixture that increases concrete's workability. Also called water reducer because it can be used to reduce water content whilst keeping workability the same.

plate dowel – steel plate, normally square or rectangular in shape, used to transfer load across a joint.

poker vibrator – tool for compacting concrete by vibration and meant to be inserted into the wet concrete. Also called immersion vibrator or internal vibrator.

popout – floor surface defect caused when material within the slab expands and forces out a chip of concrete.

Portland cement – hydraulic cement (q.v.) made by heating a mixture of lime and clay and grinding it into a powder.

post-tensioning – prestressing (q.v.) method in which the prestress is applied after the concrete has set.

pozzolana – volcanic ash used as a cement replacement material.

prestressing – process by which stresses are applied to newly placed concrete to offset the stresses that will occur later from drying shrinkage, thermal contraction and applied loads.

pre-tensioning – prestressing (q.v.) method in which the prestress is applied before the concrete has set.

profiled steel decking – corrugated steel sheet used to support and reinforce a suspended concrete floor. Also called metal deck.

pulverized fuel ash – airborne residue from the burning of coal, used as a cement replacement material. Also called pfa or fly ash.

random crack – any unplanned crack, as distinct from the planned cracks that occur at induced joints.

reinforcement – material (usually steel) used with concrete to provide strength lacking in the concrete itself.

retarder – admixture that delays the setting of concrete.

roller screed – long, straight roller that spins at high speed, used for striking off.

sand-cement screed – layer of mortar placed over a concrete slab to provide a surface suitable for floorcoverings.

saw beam – straightedge used to strike off and flatten concrete. Also called saw rod.

sawcut joint – floor joint made by sawing partway through the hardened concrete.

scratch finish – very rough floor finish made by passing a rake over the surface, and used only as a base for toppings.

screed (noun) – 1. layer of mortar placed over a concrete slab to provide a surface suitable for floorcoverings. 2. concrete topping. 3. tool for striking off, as in vibrating screed and Laser Screed. 4. pipe or rail set at finish floor elevation and used as a guide in striking off.

screed (verb) – 1. to strike off. 2. to lay a topping.

segregation – process by which materials in the concrete mix tend to separate from one another, creating zones where certain particle sizes predominate.

shear strength – resistance to cracking in a direction parallel to stress.

shrinkage-compensating concrete – concrete that undergoes an early expansion to compensate for its later drying shrinkage.

silica fume – very fine powder, a by-product of making silicon alloys, used as a pozzolan. Also called microsilica.

skip float – long-handled float designed so the user can stand off to one side of the slab. Also and more usually called bull float.

slab on grade – floor slab that rests on, and is supported by, the ground. Also called ground-supported floor.

slag – short for ground granulated blastfurnace slag, a byproduct of steelmaking used as a cement replacement material. Also called ggbs.

slump – measure of concrete's workability, usually tested by filling a mould with fresh concrete, lifting the mould at once and measuring the amount by which the concrete settles.

sprinkle finish – mixture of dry aggregate and cement sprinkled on fresh concrete and worked in with floats. Also called dry shake.

square dowel – square-section steel bar used to transfer load across a joint.

steel fabric – reinforcement made of steel wires welded into a grid. Also called wire mesh or welded-wire fabric.

structural crack – crack caused by externally applied load rather than by shrinkage or settlement within the concrete.

sub-base – course of granular material, typically crushed stone, between a ground-supported floor and the subgrade.

subgrade – levelled earth on which a floor slab rests.

sulphur concrete – concrete in which molten sulphur replaces Portland cement as the binder.

superplasticizer – admixture that greatly increases concrete's workability. Also called high-range water reducer.

surface hardener – one or more salts dissolved in water, applied to hardened concrete to increase wear resistance. Also called liquid hardener.

suspended floor – floor supported by other building elements such as columns, walls or piles.

temperature steel – minimal reinforcement required by some structural codes to handle stresses caused by thermal contraction and expansion.

tendon – steel rope used to prestress concrete.

tensile strength – resistance to being pulled apart.

tie bar – short piece of steel reinforcement crossing a floor joint. Unlike dowels (which also cross floor joints), tie bars are meant to keep joints tight.

tooled joint – floor joint finished with a hand tool that rounds off the slab edge.

trowel – finishing tool with a thin, flat blade, used to smooth and densify a concrete surface.

trowel blades – thin, sharp-edged blades fitted to a power-trowel machine. Also called finish blades.

two-way floor – suspended floor designed to span in two perpendicular directions.

unbonded topping – layer of concrete placed over but not bonded to a base slab. Also called overslab.

vapour barrier – usual American term for damp-proof membrane – sheet material placed under a ground-level floor to reduce the amount of water vapour rising from the ground. Also called vapour retarder.

vapour retarder – alternative term for vapour barrier (q.v), preferred by some authorities on the grounds that the typical product retards but does not wholly block vapour transmission.

warping joint – floor joint that relieves stresses caused by a slab's warping or curling.

water–cement ratio – ratio of water to cement, by mass, in a concrete mix.

water–cementitious ratio – ratio of water to all cementitious materials, by mass, in a concrete mix. It differs from the water–cement ratio in that it includes not only cement, but also cement replacement materials such as fly ash and slag.

wet screed – ridge of wet concrete set to finish floor elevation and used a guide for striking off. Distinct from hard screed, which describes a rigid guide such as an edge form or pipe screed.

wire mesh – common American term for steel fabric.

workability – ease with which fresh concrete can be compacted and moved.

Index

Accelerators, 123, 137, 148, 162
Acid etching, 361, 362
Additions, 160, 165, 267
Admixtures, 108, 122, 123, 137, 160
   accelerators, 123, 137, 148, 162
   pigments, 165
   shrinkage reducers, 166
   superplasticizers, 112, 115, 118, 137, 165
   water reducers, 123, 164
Aggregate interlock, 140, 155, 233, 234, 240
Aggregates, 78, 152
   coarse, 118, 119, 125, 152, 342
   fine, 119, 122, 152, 157, 342
   lightweight, 120, 156
Air-bearing vehicles, 45
Alkali-aggregate reaction, 157
Asphalt, 46, 151, 351
Automated storage-and-retrieval systems, 34

Baseplates, 25, 27, 79
Batching, 174, 306
Battery-charging areas, 22, 37
Bearing walls, 84
Bitumen, 151
Bleeding, 118, 120, 150, 274, 334, 358
Block stacking, 23
Bondbreakers, 200

Broom finish, 44, 293
Bulk storage, 23
Burnished finish, 292

Calcium chloride, 163
California bearing ratio, 73
Car parks, 46, 87, 102, 103
Carbon dioxide, 344
Carbonation, 344
Carton forms, 76
Cements, 137, 141
   expansive, 149, 267
   low-heat, 148, 204
   non-hydraulic, 151
   rapid-hardening, 122, 137, 147, 148
   sulphate-resisting, 147, 353
Cement–sand screeds. See screeds
Chaplin abrasion tester, 338
Chemical attack, 45, 100, 151, 195, 214, 249, 349
Chequerboard layout, 181
Clad-rack warehouses, 21, 25, 28, 59, 82
Clamp trucks, 31
Cleanliness, 347, 352, 362
Coatings, 166, 292, 343, 350, 355
Cold stores, 47, 142, 143
Coloured concrete, 19, 166, 295, 300
Combination blades, 285, 287
Compacting factor, 110, 115
Compaction, 193, 273, 280, 306, 350

Consistence, 107
Conveyors, 30
Corrosion, 46, 102
Cover, 101, 260
Cracks, 98, 138, 140, 155, 167, 207
  control, 48, 59, 94, 253, 302
  drying-shrinkage, 210, 253
  plastic-settlement, 101, 125, 208
  plastic-shrinkage, 207
  repair, 214
  sealing, 43, 143, 214, 352
  structural, 212
  thermal-contraction, 148, 210, 211, 253
  width, 212
Crazing, 121, 195, 209
Creep, 151, 219
Crusher-run, 78
Curing, 86, 150, 151, 195, 220, 344, 350
  compounds, 199, 344
  duration, 204, 344
  efficiency index, 201
  late, 204, 346
  wet, 196
Curl, 219, 327
  connection with concrete strength, 138
  connection with drying shrinkage, 140, 155, 167
  effect of damp-proof membranes, 274, 358
  false, 125, 221
  over metal deck, 91

Damp-proof membranes, 37, 56, 77, 222, 261, 353
Damp-proofing, 16, 37, 356
Defined-traffic floors, 33, 36, 312, 314, 322
Deflection, 91, 94, 96, 103, 104, 326
Delamination, 121, 122, 162, 296, 300, 307, 344
Dock levellers, 82
Double tees, 46, 87
Dowels, 58, 188, 214, 217, 225, 232–235, 261
  plate, 238

  size, 236
  spacing, 236
  square-section, 238
Drag factor, 264
Dry shakes, 295, 342
  and bleeding, 121
  and workability, 117
  applying, 296
  coloured, 166, 295
  effect of entrained air, 162
  effect on flatness, 329
  effect on wire guidance, 36
  electrically conductive, 295
  light-reflective, 149
  slip-resistant, 44, 295
  wear-resistant, 335, 342
Drying shrinkage, 124, 140, 171
  and curing, 195
  and rapid-hardening cement, 148
  connection with concrete strength, 138
  effect of coarse aggregate, 152, 155
  effect on curl, 219
  in freezer floors, 48
  of shrinkage-compensating concrete, 268
  shrinkage-reducing admixtures, 166
Ductility, 62, 78, 128, 135
Dusting, 195

Early-entry saws, 244
Efflorescence, 166
Elastic analysis, 62
Electrical conductivity, 295
Entrained air, 122, 161, 171, 193, 296
Epoxy, 45, 46, 50, 151, 224, 239, 247, 350
  coatings, 22, 37, 343, 357, 361
  toppings, 37, 335, 357, 361
Expansive clay, 76

Factories, 41
  loading, 42

Fibres, 167
  and ductility, 62, 78, 128, 135
  compatibility with laser
    screeds, 284
  effect on concrete strength, 78
  effect on plastic settlement, 125
  effect on wire guidance, 36
  for crack control, 210, 262, 266
  for control of thermal
    contraction, 143
  for structural reinforcement, 59, 62
  in jointless floors, 262
  in metal-deck slabs, 91
  in pile-supported slabs, 93
  length, 262
  workability test for, 110, 116
Fineness modulus, 159
Finishability, 108, 117, 152
Finishing, 277, 328, 341
Finite-element analysis, 60
Flatness, 311
  at joints, 231, 319
  effect of concrete mixing, 173
  effect of curl, 220
  effect of laser screeds, 189, 283
  effect of plastic settlement, 125
  effect of side forms, 103
  effect of superplasticizers, 165
  F-numbers, 313
  for floorcoverings, 362
  straightedge tolerances, 325
  suspended floors, 320
  TR 34, 320
Flat-plate floors, 46, 88
Flat-slab floors, 89
Float finish, 293
Floating, 278, 284, 293
Flow-table test, 110, 115
Fly ash, 122, 137, 146, 149,
    150, 204
Forklift trucks, 31, 64
Forms, 90, 93, 190, 231, 327
  carton, 76, 77
  effect on flatness and
    levelness, 103
Freezers, 47, 142, 143
Frost damage, 162, 195
Frost heave, 47
Furan grout, 352

Garages, 14
Glues, 216
Gradation of aggregates, 152, 153, 158,
    171, 208, 220
Grading, 80, 211, 273
Grinding, 290, 294
  construction joints, 231, 363
  early-age, 44, 294
  for floorcoverings, 361, 363
  for slip resistance, 44
  freezers, 48
  superflat floors, 331
  television studios, 51
  terrazzo, 309
  to improve wear resistance, 345
  to repair crazing, 210
  to repair curled slabs, 224
Ground-supported floors, 55

Hardeners, 201, 335, 343, 346
Hollowcore planks, 87, 102
Hospitals, 18
Hot-poured sealants, 251
Houses. See Residential floors
Hybrid stackers, 35

Ice, 160
Impact resistance, 138
Inclinometers, 331
Institutional floors, 18
Insulation, 47, 48, 304
Inverted slump cone, 110, 116

Jointless floors, 258
Joints, 227
  armoured, 249
  construction, 230
  contraction, 232
  depth, 242, 244, 248
  expansion, 231, 238, 245
  flatness, 319
  filling, 43, 51, 140, 143, 245
  in toppings, 302

Joints (Continued)
  isolation, 229, 245, 273
  layout, 223, 254
  loose, 140, 142
  sawn, 91, 98, 243
  sealing, 42, 249, 352
  spacing, 141, 254, 264
  types, 228
  warping, 233, 235
Joists, 85, 91, 98

Keyways, 233, 234, 241
K-slump, 110, 113

Large-bay construction, 185, 268, 279
Laser screeds, 188, 283
  as alternative to flowing concrete, 165
  effect on flatness and levelness, 315, 328
  for compaction, 194
  history, 9
  on defined-traffic floors, 315
  on reinforced slabs, 261
  on unreinforced slabs, 254
  replacing narrow-strip construction, 183
  with highway straightedges, 280
Lean concrete, 56, 78, 79
Levelness, 104, 311
  effect of laser screeds, 189, 283
  effect of side forms, 103
  F-numbers, 313
  suspended floors, 320
  TR 34, 320
Libraries, 18, 142
Lightweight concrete, 120, 129, 153
Liquid hardeners, 201
Load
  classes, 37
  safety factors, 68, 94
  transfer, 63, 142, 152, 167, 188, 223, 224, 232, 233, 262, 263, 266
Loading, 66, 81, 98
  area, 70
  position, 71

Mat foundations, 82
Metal-deck floors, 90, 98, 103, 140, 219
Mezzanines, 21, 27
Microsilica, 46, 150
Mix design, 121, 169, 211, 305
Mixing, 169, 172
Mobile racks, 28
Modulus of elasticity, 130, 139
Modulus of rupture, 58, 60, 131
Modulus of subgrade reaction, 64, 73
Moisture content, 360
Monomolecular film, 208
Moveable partitions, 17
Mud slabs, 47

Narrow-strip layout, 182, 235, 265, 279, 297, 328

Office floors, 17
One-way floors, 84, 98
Order pickers, 34

Pallet transporters, 30
Pans, 285, 287, 296, 328
Parking garages, 46, 87, 102, 103
Performance specifications, 3, 7, 140, 334, 358
Pigments, 165, 309
Piles, 21, 55, 74, 76, 83, 92, 99
Plastic analysis, 62
Plastic settlement, 51, 101, 124, 155, 193
Plastic shrinkage, 152
Polysulphide sealants, 251
Polyurea, 51, 224, 239, 247
Polyurethane sealants, 251
Popouts, 157
Portal frames, 81, 229
Post-tensioning, 60, 63, 102, 268
  and need for compaction, 194
  car parks, 46
  effect on curl, 221
  effect on wire guidance, 36
  for air-bearing vehicles, 45

Post-tensioning *(Continued)*
   for crack control, 43, 45, 143, 210, 268
   for structural strength, 79
   freezer floors, 48
   in car parks, 46
   isolation joints, 226
   joint opening, 234
   over poor ground, 63, 74, 76
   pile-supported slabs, 93
   testing for, 138
Power trowels, 152, 285, 287
Pozzolans, 137, 149
Precast floors, 86, 94
Prestressing, 102, 267
Pre-tensioning, 86, 102, 210
Profileographs, 330
Pumping, 112, 113, 117, 155, 179

Racks, 24, 64, 68, 93
Rack-supported buildings, 21, 25, 28
Raft foundations, 82
Random-traffic floors, 312, 315, 323
Reach trucks, 31
Ready-mix concrete, 172, 306
Reinforcement, 100, 256, 257
   at joints, 234
   at re-entrant corners, 256
   continuous, 43, 48, 58, 258
   effect on curl, 222
   effect on wire guidance, 36
   for crack control, 94, 143, 257
   for stable joints, 263
   in metal-deck slabs, 91
   in pile-supported slabs, 93
   nominal, 263
   over poor ground, 74
   wire mesh, 59, 91, 93, 101, 183, 265
Residential floors, 14
Retaining walls, 82, 229
Retarders, 123, 164
Roller screeds, 282, 328
R-value, 49

Safety factors, 68, 94
Sand, 56, 159, 342

Sand–cement screeds. See screeds
Scabbling, 361
Scarifiers, 361
Schools, 18
Scratch finish, 294
Screeds, 303
   floating, 51, 304
   for flatness, 320, 363
   in television studios, 51
   mixes for, 170, 174
   moisture in, 360
   testing, 307
   with damp-proof membranes, 358
Semi-rigid joint fillers, 43, 215, 224, 239, 246
Serviceability, 94
Setting time, 122
Settlement, 74, 93
Shear, 89, 94, 96, 100, 135
Shelving, 26
Shot-blasting, 44, 297, 361
Shrinkage-compensating concrete, 43, 58, 143, 149, 210, 221, 267
Shrinkage-reducing admixtures, 166
Sieve analysis, 153
Silica fume, 146, 150
Silicone sealants, 251
Slag, 122, 137, 146, 149, 150, 204
Slipsheets, 58, 211, 261, 358
Slump, 107, 110, 171
   and entrapped air, 193
   and finishability, 118
   classes, 112
   effect of pumping on, 180
   for dry shakes, 296
   for superflat floors, 331
   testing, 110, 180
Soil cement, 56
Span, 99
Speculative warehouses, 5, 38
Stacker cranes, 34
Straightedges, 183, 186, 190, 278, 328, 331
   highway, 121, 279, 331
   testing with, 307, 311, 325
Strength, 78, 100, 127, 138
   characteristic, 136
   compressive, 64, 128, 171

Strength *(Continued)*
  flexural, 50, 58, 60, 64, 78, 95, 130, 131, 171
  gain, 137, 148, 150
  shear, 96, 100, 130, 135
  tensile, 101, 133, 210, 212
Structural design, 5, 55
Sub-base, 24, 56, 74, 75, 77, 80
Subgrade, 24, 55, 73, 80, 213
  modulus, 64, 73, 75
Sub-slab friction, 261, 265, 267, 274
Sulphur concrete, 151, 350
Superflat floors, 329
  ACI classification, 336
  construction methods for, 328, 329
  continuously reinforced, 258
  effect of curl, 221
  effect of plastic settlement, 125
  forms for, 190
  in television studios, 51
  specifying, 315
  straightedges for, 279, 280
Superplasticizers, 112, 115, 118, 137, 165, 193

Television studios, 50
Terrazzo, 19, 309
Texture, 44
Thermal contraction, 36, 47, 85, 142, 211, 215, 219
Tie bars, 188, 233, 235
Tiles, 351, 357
Toppings, 7, 299, 346, 350, 355
  bonded, 301, 361
  chemical-resistant, 350
  curling of, 140, 219
  epoxy concrete, 151, 350
  flat, 320
  high-strength concrete for, 129
  monolithic, 299
  on precast floors, 87

  preparation for, 292, 294, 301, 355
  repairing, 224
  site mixing for, 173
  small batches for, 177
  unbonded, 302
  wear-resistant, 335, 346
Transporting concrete, 175
Trowelling, 278, 286
Two-way floors, 88
Turret trucks, 32, 311, 313
Tyre marks, 199

Unreinforced floors, 58, 63, 163, 254, 284

Vacuum dewatering, 121, 296
Vebe test, 110, 115
Very-narrow-aisle trucks, 32, 311, 315
Vibrating screeds, 183, 190, 194, 280, 328
Vibration, 116, 183, 193, 281, 306
VNA. See Very-narrow-aisle trucks

Waffle floors, 89, 98
Warehouses, 5, 21, 249, 268
Water reducers, 123, 164
Wear resistance, 44, 100, 121, 130, 151, 158, 333
  dry shakes, 295
  effect of curing, 195, 204
  effect of vacuum dewatering, 297
  testing, 336, 337
  toppings, 299, 336
Wet screeding, 186, 279, 306
Wide-strip layout, 184
Wire-guided vehicles, 35, 101
Workability, 107, 152, 180, 193